Scientific Poultry Production and Nutrition

Scientific Poultry Production and Nutrition

Michael Youn
Editor

KOROS PRESS LIMITED
London, UK

Scientific Poultry Production and Nutrition

© 2012

Printed in 2017 for Sale in the Indian Subcontinent

Published by
Koros Press Limited
3 The Pines, Rubery B45 9FF, Rednal,
Birmingham, United Kingdom

Tel.: +44-7826-930152
Email: info@korospress.com
www.korospress.com

ISBN: 978-1-78163-090-7

Editor: Michael Youn

10 9 8 7 6 5 4 3 2 1

British Library Cataloguing in Publication Data
A CIP record for this book is available from the British Library

Exclusively distributed by CBS Publishers & Distributors Pvt. Ltd.
Sales & Distribution Rights only for India, Pakistan, Bangladesh, Sri Lanka, Nepal and Bhutan.This book is not to be sold outside these territories.

Contents

Preface

Poultry farming is the practice of raising domesticated birds such as chickens, turkeys, ducks, and geese, as a subcategory of animal husbandry, for the purpose of farming meat or eggs for food. More than 50 billion chickens are raised annually as a source of food, for both their meat and their eggs. Chickens raised for meat are called broilers, whilst those raised for eggs are called laying hens. In total, the UK alone consumes over 29 million eggs per day. Some hens can produce over 300 eggs a year. Chickens will naturally live for 6 or more years. After 12 months, the hen's productivity will start to decline. This is when most commercial laying hens are slaughtered.

The majority of poultry are raised using intensive farming techniques. According to the Worldwatch Institute, 74 percent of the world's poultry meat, and 68 percent of eggs are produced this way. One alternative to intensive poultry farming is free range farming. Friction between these two main methods has led to long term issues of ethical consumerism. Opponents of intensive farming argue that it harms the environment and creates health risks, as well as abusing the animals themselves. Advocates of intensive farming say that their highly efficient systems save land and food resources due to increased productivity, stating that the animals are looked after in state-of-the-art environmentally controlled facilities. A few countries have banned cage system housing, including Sweden and Switzerland. Consumers can still purchase lower cost eggs from other countries' intensive poultry farms.

Poultry breeding remains largely based on classical quantitative genetics. In essence, pedigree broiler candidates are full-fed nutritionally-dense and properly balanced diets to allow individuals that have the greatest potential to utilise crude protein (CP) and metabolisable energy (ME) to grow fast, convert feed efficiently, and yield well to become apparent by their performance. Thus, broiler strains are often selected on high-protein, high-energy diets.

Selection on nutrient dense diets apparently necessitates nutrient-dense diets in order for the progeny to fully express their genetic potential. An excellent example of the relationship between genetic progress and appropriate nutritional compensations can be taken from research with quail. Random-bred Japanese quail were placed on a selection programme intended to create heavy weight (HW) quail. These quail were full-fed 28% CP diets for 28 days and then the largest birds were selected and mated to produce the next generation. When these birds were reared to sexual maturity on a 24% CP diet, as recommended by the National Research Council, there was an obvious delay in sexual maturity.

A useful encyclopaedia for those who are interested in the study of this subject. It may equally be liked by the general readers interested in studying the topic as also the academics and practitioners.

—Editor

Introduction

Managing Today's Broiler Breeder Female

Over the past few decades, broiler breeders have undergone intensive selection for faster growth rate, increased yield and improved feed conversion. Although these traits are measured at the broiler level, they impact the breeder hen in ways we often do not consider. The objective with broiler breeders is to have them consume an "ideal" amount of nutrients within a given time period to produce a bird whose weight, body condition and frame allow the reproductive organs to mature and function at their best. How do we combine art and science to manage the sexual maturation of today's broiler breeder female?

Photostimulation

One of the most critical time periods in broiler breeder hen management is the time from photostimulation (lighting) to peak production (Robinson, 1995). This period is characterised by relatively fast weight gains, in addition to changes brought about by the development of a functioning, hormone-producing ovary. Lighting the breeder pullet flock is generally considered the cue to initiate puberty, although the response to lighting can be modified by the feeding programme.

At photostimulation, light energy passes through the skull of the breeder pullet into the brain and "illuminates" the hypothalamus. The hypothalamus in the brain is much like the main circuit breaker in a house; it controls a variety of body processes including reproduction. The brain acts in concert with the liver, skeletal system, ovary and oviduct to make up the reproductive system in the breeder hen. After the hypothalamus receives a photostimulatory signal (long day length above a certain threshold of intensity), the hypothalamus secretes specific hormones that travel to the anterior pituitary portion of the brain (Robinson, 1999). The anterior pituitary produces hormones known as Luteinising hormone and Follicle Stimulating hormone that travel to specific tissues in the ovary to stimulate ovarian function.

One of the first responses seen when looking at the ovary of the pullet after lighting, is that the tiny ovarian follicles begin to increase in size. These small follicles produce large quantities of estrogens. Estrogen causes most of the reproductive transformation associated with puberty.

Firstly, estrogen increases the production of yolk precursors in the liver of the bird. Visibly, the liver can be seen to enlarge and become paler as it increases in fat content for production of egg yolk lipids. Secondly, the oviduct increases in size, as it must be ready to receive ovulated follicles by the time the ovary has mature follicles ready to ovulate. Thirdly, estrogen results in changes to bone composition, so that calcium can be mobilised daily to facilitate egg shell formation. Finally, estrogen, together with male sex hormones, results in changes to plumage, comb size and sexual receptivity to males. Traditionally, flocks receive photo stimulation when they are 20-22 weeks of age resulting in onset of egg production at approximately 24-25 weeks of age. This programme tends to maximise egg numbers, but may result in eggs that are smaller than standard early in the laying cycle. It also often results in egg production before hens are capable of producing a quality germ cell. Lighting birds later than 20-22 weeks allows females to become larger and more mature at the onset of production. Unfortunately, lighting birds later will likely also delay egg production until 25-26 weeks. However, this may or may not affect the total number of hatching eggs produced.

Ovulatory Cycle

Yolk is deposited into follicles as they proceed through the hierarchy to become mature. Two requirements must be met for the follicle to ovulate. First, the follicle must send a hormonal signal to the hypothalamus through the release of progesterone that signals that it is mature. Second, the hypothalamus must receive the signal from the mature follicle during a 6 to 8 hour period of the day in which the hypothalamus is responsive to the progesterone signal (Robinson, 1999). Hens that have slow rates of follicular maturation (26-28 hours or more) lay short (2-3 day) sequences. On the other hand, hens that lay very long sequences typically have maturation rates of 24 hours, or perhaps less. Sequence length changes throughout the egg production year with the longest sequences seen at the time of peak production at about 30-35 weeks of age. All hens lay one characteristically long sequence of eggs known as the "prime sequence" which in broiler breeders is usually about 20 eggs in length.

Feed Requirements

While feeding programmes differ across the country due to differences in integrators, complexes, weather conditions, seasons and genetic strains of birds, it is important to be continually adjusting the feeding programme to provide the nutrients needed for optimum performance. Breeders require these nutrients for body maintenance, growth and egg production.

Body maintenance requirements, which include maintaining body temperature and systems within the bird that allow for digestion, respiration, excretion and immune response, range from 50 to 75% of a hen's daily needs. As with most animals, body maintenance needs have priority, since the breeder hen must maintain her own body to survive. While the growth needs of hens during the post-peak production period do not contribute greatly to the hen's daily nutrient requirements, pre-peak growth can be substantial.

Nutrient needs for reproduction are a function of the number and size of eggs produced. In general, egg production exerts more influence on nutrient requirements than does egg size. This is part of the reason a service technician always has his/her calculator in hand and adjusts the feed allocation on each visit to the farm. This is an attempt to maximise egg numbers and keep hen body weight on target, since overweight hens produce fewer eggs than trimmer hens.

Table 1: Times of oviposition for individual hens laying 2- to 7-egg sequences[1].

Sequence Length	Day 1	Day 2	Day 3	Time of Oviposition Day 4	Day 5	Day 6	Day 7
2 eggs	09:28 AM	01:30 PM					
3 eggs	08:08 AM	11:26 AM	02:40 PM				
4 eggs	08:20 AM	09:45 AM	01:45 PM	03:37 PM			
5 eggs	07:56 AM	09:03 AM	10:45 AM	01:11 PM	03:05 PM		
6 eggs	07:20 AM	07:59 AM	09:04 AM	10:11 AM	12:56 PM	03:40 PM	
7 eggs	07:47 AM	08:15 AM	09:20 AM	09:40 AM	11:36 AM	01:09 PM	03:24 PM

[1] Adapted from Robinson, 1999.

Flock Uniformity

Flock uniformity is critical to proper feed allotments. If there is a great deal of variability in body weight, and all birds have equal opportunity to eat, the small birds will over-consume and larger birds will under-consume in relation to their nutrient requirements (Robinson, 1999).

Uniformity issues are most critical at the time of photo stimulation and will usually result in poor peak performance as well as significant problems in post peak periods. In non-uniform flocks, birds receive the same feed allotment, but feeds are formulated for birds in lay. Since birds in lay have higher nutrient requirements than non-laying birds, nonlaying birds will over consume relative to their requirements and get fat, which will hinder future performance. Clearly, uniformity is necessary to obtain peak performance in breeder females.

Summary

Properly managing the sexual maturation of the modern broiler breeder female is critical to obtaining a high peak and large overall number of quality hatching eggs. The most critical management period for broiler breeders is from photo stimulation (lighting) to peak production. Management deficiencies during this period are always costly and often cannot be compensated for at a later date.

Broiler breeders require nutrients for maintenance, growth and egg production. Maintenance needs are met first and until that happens, growth and egg production are reduced. Adjusting the feed allotment throughout the lay cycle controls bird nutrient intake. Intake must be strictly controlled to prevent hens from becoming overweight resulting in decreased egg production.

Flocks must be uniform in weight and body condition in order to properly allocate feed allotments. Uniformity is especially critical at the time of lighting. Flocks that vary excessively in uniformity are nearly impossible to properly manage from a feed allotment standpoint. This will have a negative impact on performance and may lead to a low, flat peak and decreased overall production. Remember that the key to managing the modern broiler breeder female is a combination of 1) correct body weight and uniformity, 2) light stimulation, and 3) feed stimulation. A sound, consistent management programme must be in place that will address each of these areas in order to be successful.

Broiler Breeders

A Farm Animal Welfare Network Factsheet

'We cannot support your suggestion that parent stock birds are badly exploited; rather, to the contrary, they are reared carefully and sensitively and thereafter live long and healthy lives.'

FAWN argues that the broiler chickens which make up the breeding stock suffer both mentally and physically. Bred to be 'greedy', the parent birds must be kept for extended periods on severely restricted rations if they are to survive and reproduce successfully. Through genetic selection the modern poultry industry has bred a bird with inherent health and welfare problems.

How many breeders are there in the UK?

Approximately seven million, supplying around 700 million fertile eggs annually. Eighty to 90% of these birds are female.

How are they Housed?

In the UK breeders are housed intensively, in windowless controlled environment sheds, usually holding several thousand birds per unit. There are no laws or MAFF codes of recommendations relating to stocking density for breeders, but they are not so closely stocked as young broilers.

Lighting must be bright enough to encourage laying, therefore it is impossible to control aggression in males via low lighting regimes. FAWN has been told of significant aggression among cockerels in breeding units, including aggression towards stockpeople. FAWN has been informed, by a reliable source, that some stockpeople feel the need to 'show who is boss' in ways which could lead to prosecutions for cruelty if their actions were reported. The birds are kept on litter (wood shavings or chopped straw) and sheds contain nesting boxes. Some litter is eaten by the hungry birds.

The sexes are kept at a ratio of nine birds to one cockerel. Hours of 'daylight' (in fact artificial light) are gradually increased from eight hours in the early weeks to 16 hours.

How do the parent birds differ from the young broiler chickens we see in the supermarket?

Putting it simply, the modern broiler chicken has been selectively bred over the last five decades to consume large amounts of high protein food, and to put on weight rapidly. This process has been exaggerated by lighting regimes (to encourage birds to eat throughout the 24 hours) and by adding growth promoters to feed. The poultry industry has been so successful in its quest for fast-growing chickens that today's bird can weigh up to four times that of slower growing strains.

The Parent Stock/Breeders

The parent birds are just as 'greedy' as those slaughtered at 6-7 weeks (i.e. those that end up on the supermarket shelf, etc.) but if fed to appetite many would die pre-maturely, from diseases associated with obesity (e.g. heart failure). Fertility would be dramatically reduced, since overweight males would have difficulty mounting the females, and females would suffer from a high incidence of multiple ovulations, and soft shelled eggs, all useless for hatching purposes. Joint, bone and feet problems would also increase. All these conditions combined would result in an uneconomic breeding stock.

Can chickens really be 'greedy'?

The modern broiler chicken is 'greedy', though the bird should not be blamed! Dr Mench of Maryland's Department of Poultry Science explains the 'greediness' as follows:

> *'The selection of broilers for increased growth rate has resulted in an increase in appetite (Siegel and Wisman, 1966) by modulating both central and peripheral mechanisms of hunger regulation (Lacy et al., 1985, Denbow 1989). The increased food intake causes obesity, which must be controlled in broiler parent stock in order to maintain reproductive competence.'*

How has the poultry industry 'solved' the problem of over-eating by parent stock birds?

Both sexes being reared as parent stock are fed on severely restricted rations for most of their growing period (2-20 weeks, approximately). Males continue to be fed on severely restricted diets throughout their lives, females more moderately, since they must produce eggs.

To what degree is feed restriction practised on female breeders?

For females, who must produce eggs, the restriction is less severe. For example, a female breeder will receive 52 g of feed daily at 7 weeks of age, while a 'supermarket' female of the same age will consume 182 g of feed. The amount of feed offered goes up gradually to reach 168 g daily by week 26-27, and is maintained at this level until slaughter (at around 60 weeks of age).

How does the female breeder compare in weight with a 'supermarket' female bird?

At seven weeks of age a breeder weighs 1.7 lb (780 g) and a 'supermarket' chicken (before processing) weighs around 5.38 lb (2440 g).

The female breeder's ultimate weight is around 7.2 lb (3260 g). When fully-grown, she will be consuming less feed than a 'supermarket' female chicken of seven weeks.

How severely restricted is the breeding cockerel's feed?

At seven weeks old, a 'supermarket' male broiler (before processing) weighs 6.39 lb (2897 g) and will consume 205 g feed daily. His breeding counterpart of the same age will weigh 2.4 lb (1100 g) and consume 78 g of feed daily - a huge reduction on what he would eat if fed ad libitum.

How much is a breeding male allowed to consume when fully grown?

On restricted feed, a mature cockerel weighs around 10 lb (4550 g) and consumes 120 g of feed daily.

How much would a 'supermarket' cockerel consume?

Sometimes, in the run-up to Christmas, broiler chickens are 'grown on' beyond the usual 6 or 7 weeks (the age at which most are slaughtered). A 'supermarket' male bird weighing 10 lb at 10 weeks of age consumes 253 g of feed daily.

Figures in facts 13-18 are based on information published by The Cobb Breeding Company. Ross Breeders, UK, published performance objectives in 1995 which include the following figures:

Gender	Feed consumption at 7 weeks	Weight at
	g/bird/day	7 weeks
Female	53g	760g
Male	72g	1100g

NB Most 'supermarket' chickens are slaughtered at seven weeks, or earlier.

How do the birds react to feed restriction?

'These rations are usually provided daily from the second week of life onwards, according to programmes recommended by the commercial breeders, and amount to a reduction in consumption of some 55-75% compared with birds of the same age fed ad libitum. In contrast to ad libitum feeding, birds that are fed on such rations tend to be very active, and many of them spend much of their time pecking in a stereotyped manner at various non-feed objects (in the absence of food - Ed.), such as drinkers or pen walls.' (Kostal, Lubor et al., (1992) 'Diurnal and Individual Variation in Behaviour of Restricted-fed Broiler Breeders', Applied Animal Behaviour Science, 32: 361-374)

To sum up:

> *'Breeders are (therefore) fed only sufficient during rearing*
> *to reach about 50% of their potential body weight.'*
>
> > *(Poultry Production Systems, Appleby,*
> > *Hughes and Elson, CAB 1992)*

NB Eventually, breeding chickens achieve roughly two thirds of their potential bodyweight.

How are broiler breeders' rations given to them?

Up to 20 weeks of age (approximately) the birds are generally reared separately. This makes it easier for the farmer/company to administer the approved amount of feed to males and females. At this stage there are two methods for distributing the feed - either it may be scattered (in pellet form) on the shed floor, or fed in troughs. The first method is preferable, as it encourages birds to scratch around in the litter, thus keeping it in good condition. It also makes the birds' life more interesting, since feed will take longer to find, and aggression connected with competition at the feed troughs is minimised.

What happens when the sexes are put together, for breeding?

At this stage (around 20 weeks) although the birds are together, their feed is usually separate.

> *'As the potential growth rate of the breed improves with*
> *each new generation, it is even more important to control*
> *the bodyweight of the parent stock. The basic principle*
> *of separate sex feeding is to exclude the males from the*

> *female feed track and provide a separate male feeding
> system. The normal method of exclusion is a grill placed
> on top of the track which controls access by the horizontal
> distance between the bars.'*
>
> *(Cobb 500 Breeder Management Guide).*

Whatever method is used, the principle is to make access to the females' feed too narrow for the males' (bigger) heads. The males' feed is suspended out of reach of the females.

Daily Feeding

In the UK the law states that livestock must be fed every day. The 1994 Welfare of Livestock Regulations state:

'Livestock shall be provided with an adequate supply of fresh drinking water each day and shall have access to food each day, except where a veterinary surgeon acting in the exercise of his profession otherwise directs.'

A spokesperson for the UK Ministry of Agriculture has informed FAWN that MAFF does not know whether the Schedule's requirements apply in other EU countries. Skip-a-day programmes are practised widely in the USA and probably in many countries worldwide, perhaps including members of the EU, hopefully with the exception of the UK.

Skip-a-day Feeding

Feed is given every other day. No feed is given on the 'skip' days.

Five Days/Week Feeding

The Cobb Breeding Company describes this as a 'compromise' between everyday feeding (which it recommends) and skip-a-day feeding. The system involves omitting all feed on two days out of every seven. A spokesman for the Cobb Breeding Company explained to FAWN that in suggesting the five days/week programme, the company is trying to encourage farmers in the direction of daily feeding. Ross Breeders state: 'Every day feeding is preferred whenever volume permits'. This company also describes systems which omit 'complete feed' on one or two days a week but include a scratch feed on the 'no feed' days.

Why Skip-a-day?

Broiler breeders' rations are so restricted that birds are extremely hungry for long periods. In units without sufficient feeder space it was found that some birds rushed to feeders and, in the overcrowded conditions, consumed more than their 'fair share', leaving others truly starving. One 'solution' to this problem is to give twice the birds' daily ration every other day, on the assumption that during the time taken to eat 'double rations' all birds will have the chance to get enough to eat to ensure survival.

Which system involves the least suffering?

FAWN believes that the severe feed restriction practised throughout the industry, for boiler breeders, is unacceptable in welfare terms, but clearly skip-a-day and five days/week programmes represent the worst options. Stress, from chronic hunger, is clearly present in all three systems:

> 'Broiler breeders are reared on feed restriction from about two to 18 weeks to avoid them becoming too fat for laying. They are therefore under considerable stress, especially at about 12 weeks of age.'

> (Mark Pattison, MRCVS, writing in Poultry Practice, Ed. Edward Boden, Bailliere Tindall, 1993).

When is the feed given?

Usually in the early morning - birds finish their daily ration within 10-15 minutes.

How do the birds react to rationing?

'Despite its positive influence on health and reproduction, there is mounting evidence that food restriction has a negative effect on welfare. Fowls normally spend a considerable portion of their day in activities associated with foraging, and when given a choice prefer to work for at least part of their daily intake of food rather than eating it all from a free supply. Food restricted broiler breeders, however, consume their food ration in a very brief period of time. They may also be chronically hungry, since their level of motivation to consume food is approximately four times that of ad libitum-fed birds subjected to 27 hours of food withdrawal.. Food restricted broiler breeders show behaviour which is indicative of frustration of feeding motivation. Restricted males are more aggressive than fully-fed males, while restricted hens and pullets are more fearful and active and also display high rates of pecking stereotypes. Overdrinking is also a common problem in broiler breeder flocks, resulting in the need to restrict water intake as well as food intake in order to maintain litter quality.'

NB Overdrinking occurs when birds are chronically hungry, by way of compensation. (It is illegal to withhold feed for one day on UK farms, but establishments carrying out research on farm livestock are not answerable to farm animals legislation, but operate under Home Office Regulations.) A behavioural change brought about by hunger is highlighted by Ross Breeders:

'It is unlikely that male body weight can be maintained on less than 125g/bird/day. If less feed than this is allocated to the male feeder then the males will take feed from the female feeder to maintain an adequate intake or, if excluded from the female feeder, will begin to lose weight. Males stealing female feed can significantly reduce egg numbers.'

Two points emerge from this: a) hungry males may damage their heads, in attempts to get through narrow grids and b) cockerel behaviour has been grossly degraded by modern commercial conditions, as anyone who has witnessed cockerels' normal behaviour* will agree. *Under natural conditions cockerels will seek out food for female birds, drawing their attention to desirable titbits, and standing back to allow the hens to eat first. A spokesperson for the Cobb Breeding Company informed FAWN that cockerels' 'gentlemanly' behaviour can still be observed in breeding sheds.

Is physical damage caused by feed restriction?

Broiler breeders are culled when around sixty weeks of age, so long-term effects of severe feed restriction may not be known. Injuries may be caused, indirectly, by the birds' hungry state:

> '(Staphylococcal arthritis) is caused by Staphylococcus aureus which invades the tissue or blood stream, following injury to the skin, especially the feet. Any environmental factor which may result in skin injury, e.g. sharp projections, wood splinters in litter, or birds suffering injury when rushing to the feeders where feed restriction is practised (in broiler breeders particularly) will result in an increased incidence of this condition.'
>
> (G.S. Coutts, BVMS MRCVS, Poultry Diseases Under Modern Management).

> '*At this time (twelve weeks of age) staphylococcal infection of the hock joint is an important condition (in broiler breeders) and can cause loss through culling. Treatment with broad spectrum antibiotics is sometimes effective, but must be started early and continued for two or three weeks. This hock condition can be almost eliminated by floor feeding without a trough (Dutchman type) feeder, on which birds appear to damage themselves. Traumatic injury, often not visible, seems to set off the staphylococcal infection.*'

Poor feed trough design can result in severe damage to female broilers' heads.

> '*In broiler breeding flocks, separate sex feeding is being widely adopted as a means of controlling bodyweight gain in male birds with the objective of improving fertility. In breeding pens, grids fitted to the feeding trough allow females to feed, but males, which have wider heads, cannot gain access. From 45 weeks of age, a proportion of female birds were observed to have swollen heads, but only in that group fed from troughs fitted with grids. At 60 weeks of age, approximately 15 per cent of birds from that group were affected. In this flock, the head swelling was judged to be of traumatic origin and a consequence of fitting grids of incorrect size to feeding troughs.*'

> *(S.R.I. Duff et al., Head Swelling of Traumatic Origin in Broiler Breeding Fowl, Veterinary Record, August 5th 1989).*

NB Before this research, Swollen Head Syndrome, a viral disease, had often been confused with the damage caused by too-narrow grids. A photograph accompanying shows a bird with severe facial/head swelling and closed eyes. A leading UK broiler breeding company has assured FAWN that 'lessons have been learned' and grid dimensions improved, but it must be a cause for concern that, worldwide, badly-designed troughs may still be in use, causing horrific suffering. Bearing in mind the date of the VR (1989) it seems likely that 'old-style' equipment may not have been replaced in many broiler breeding units, where troughs, etc. might be expected to have a life of a decade or two.

Clearly, much suffering has been/is being caused to broiler breeders through trough feeding of scant rations. Floor scattering of pelleted food is much more humane, but does not answer the demands of single sex feeding, which is usually considered necessary once the sexes are put together in the same building.

Are broiler breeders beaktrimmed?

A major breeding company recommends beak trimming at a day-old or at 4-5 days old should this mutilation be deemed necessary because of past problems (head damage etc).

'Beak trimming of both males and females is not recommended unless there has been a history of physical damage or when it is clear that more suffering would be caused in the flock if it were not carried out, (e.g. mating behaviour can sometimes be the cause of damage which may result in subsequent infection particularly to the head)'.

NB The chick and adult chicken depicted on the manual's cover are both beaktrimmed/debeaked. The terms 'beaktrimming', 'debeaking' and 'partial beak amputation (PBA)' all describe the process of removing a proportion of the bird's beak, to minimise damage from aggression.

Are broiler breeders mutilated in other ways?

Male broiler breeders are despurred:

> 'To avoid injury to hens during mating, the last joint of the inside toes of male breeding birds may be removed. This must be done within the first 72 hours of life. A veterinary surgeon must carry out the operation if it is performed after the first 72 days of life.'
>
> *(Codes of Recommendations for the Welfare of Livestock, issued by MAFF, and the Secretaries of State for Scotland and Wales, dated 1987, para. 51).*
>
> 'Acute necrotic enteritis was diagnosed by Dumfries (State Veterinary Centre - Ed.) in six-week-old broiler breeder cockerels which were recovering from an earlier episode of staphylococcal femoral necrosis associated with despurring.'
>
> *(Veterinary Record, October 22nd 1994).*

Frequent Matings

'Optimum mating' by males can be judged by the vent colour - bright red indicates the most 'keen' (and therefore profitable) birds.

Cockerels displaying paler vents are removed from the flock, and culled, on a regular basis. Overmating of females can cause severe feather loss, resulting in scratched and torn skin.

Apart from being chronically hungry, aren't the breeders better off than their offspring, since the parents don't suffer from diseases associated with obesity?

Male breeders may still experience chronic orthopaedic problems which can cause pain.

In addition to orthopaedic (hip and leg, etc.) problems, broiler breeders also suffer from many of the viral diseases common in young chickens and respiratory diseases.

For how long do they live?

Most broiler breeders are slaughtered when about sixty weeks old, once past their peak of reproductive powers.

What about catching and slaughter?

Males are heavy (up to 10 lb - Ed.) aggressive and very active, making them extremely difficult (and tiring) to catch. Orthopaedic problems can make shackling difficult and probably painful.

Animal Aid has produced a video entitled 'Here's the Catch' which reveals the horrific cruelty meted out to young broilers at the time of catching. Heavier birds (breeders) are likely to suffer even more at this time, as catchers carry birds by one leg (and more than one in each hand - 4-5 in each hand is routine when collecting younger birds). NB 'Here's the Catch' is available from Animal Aid and FAWN.

Slaughter represents another area of suffering, since the conditions outlined in this fact sheet will ensure that acute pain is felt by many birds as they hang in shackles, awaiting slaughter.

What happens to broiler breeders, finally?

They are processed into pies, canned soups, etc. Has there been an enquiry into the welfare of broiler breeders, in the UK?

The House of Commons' Agriculture Committee's second Report into the UK poultry industry makes no mention of the broiler industry's breeding stock under its Welfare Recommendations.

The Farm Animal Welfare Council (FAWC), the government appointed body which reports on farm animal welfare, has investigated the broiler industry. Its Report on the Welfare of Broiler Chickens

(April 1992) omits mention of the breeding stock, since the Council felt the problems associated with parent birds are so distinct from those found in younger birds. (However, in its January 1992 Report on the Welfare of Turkeys, the parent stock are included.) At the time of writing, FAWC is not engaged in investigations regarding the welfare of broiler breeders.

Challenge Feeding

High and sustained peak egg production can only be achieved with uniform breeder flocks fed to meet their nutritional requirements. With 85-88% peaks now possible in the industry, it is obvious that we have to carefully plan and execute a feeding programme tailored to meet the breeders' nutrient needs. Underfeeding results in very short duration peaks, of only 3-4 weeks, and those are usually associated with the classical sign of loss or stall-out in body weight for 1-2 weeks. On the other hand overfeeding, especially with energy, will result in excessive weight gain, and while peak production may be little affected, there will be precipitous loss in egg production through 34-64 weeks of age. The basis of feed allocation at this critical time is obviously to allow genetic potential for increases in both egg numbers and egg size, and also to allow for modest weekly gains in body weight. Managers should consider "challenge feeding" as part of their feed management system at this critical time.

Challenge feeding involves giving the hens extra feed on 2 or 3 days each week, based on need, without changing the base feed quantity scheduled for the flock. For example, a flock may receive 168 g/bird/day at peak, with an additional "challenge" of 7 g/bird/day given three days each week. The challenge feed is, therefore, equivalent to 3 x 7g / 7d = 3 g/bird/day. In reality birds receive the equivalent of 168 g + 3g = 171 g/bird/day. The immediate question is why bother with this more complicated system, and just give the flock a base feed allowance of 171 g/bird/day?

The advantages of challenge feeding, rather than simply increasing the base allocation are:

1. On days of challenge feeding, feeding time will increase, and this helps to improve uniformity.

2. It is easier to make adjustments to nutrient intake based on day-to-day change in needs as may occur with changes in environmental temperature.

3. Birds become accustomed to change in feed allocation, which will be important once feed withdrawal is practised after peak.

4. Ease of tailoring nutrient needs to individual flocks. For example, a base feed allocation of 165-175 g/bird/day may be standardised across all flocks, with individual flock needs at peak being tailored with the quantity and/or frequency of challenge, depending upon actual production, environmental temperature, etc.

The actual quantity and timing of challenge feeds must be flexible if they are to be used efficiently. In practice the challenge should not represent more than 5 per cent of total feed intake, and most often the quantity will be 2-4 percent. On the other hand, the quantity of the challenge should be large enough to meaningfully contribute to the factors listed previously. For this reason there needs to be a balance between the quantity of feed given, and the frequency of this feeding.

For example, a daily challenge of 3 g/bird/day given daily will be much less effective than 7 g/bird/day given 3 times each week. In both instances birds are receiving 21 g/week as a challenge, but in the later example the challenge quantity is more meaningful and we are more likely to see a bird response in terms of egg output.

Challenge feeding should start when birds are at 60-70% production, and should be discontinued when egg production falls below 80 per cent. For most flocks, therefore, we can expect to practice challenge feeding from about 29 through 40 weeks of age. The idea of challenge feeding is to more closely tailor feed allocation to breeder hen needs, and so there should be no standardised system.

Managers must be given flexibility to alter challenge feeding based on fluctuating needs. In most instances the challenge will be used to lead bids into a sustained peak. Because the concept of challenge feeding is to more closely tailor feed allocation to needs, then it is usual practice to alter the quantity and/or duration of challenge as birds progress through peak egg production.

Maximum challenge feeding should coincide with peak egg output, with lesser quantities given prior to, and after actual peak. On this basis we recommend challenge feeding to be reduced (but not discontinued) once birds are 2% below peak egg production. Following are three examples of challenge feeding tailored to three different flock situations.

High nutrient dense feed used, with good ingredient quality control. Expected high-low temperatures of 31-24 degrees Celsius. Good flock uniformity at 20 weeks of age, and previous flocks show consistent peaks of 85-87%.

Egg production	Base feed		Challenge feed	
35%	162g	(35.6 lb/100)	None	-
60%	162g	(35.6 lb/100)	5 g/d, 2 x/wk	(1.1 lb/100)
80%	162g	(35.6 lb/100)	8 g/d, 2 x/wk	(1.8 lb/100)
-2% from peak	162g	(35.6 lb/100)	5 g/d, 2 x/wk	(1.1 lb/100)
79%	162g	(35.6 lb/100)	None	-
<79%	Reduce	-	None	-

In this example, because birds are uniform in both weight and maturity and a good quality diet is used, and there is no major temperature stress, the challenge is quite mild. For this flock, a heavier challenge may result in excess weight gain. This type of mild challenge is most frequently used where feed quality is ideal, and there is minimal disease and mycotoxin exposure.

High nutrient-dense feed used, with good ingredient quality control. Expected high-low temperatures 28-14 degrees Celsius. Poor to average flock uniformity at 20 weeks of age, and previous flocks show variable peaks of 81-87%.

Egg production	Base feed		Challenge feed	
35%	162g	(35.6 lb/100)	None	-
60%	162g	(35.6 lb/100)	8 g/d, 3 x/wk	(1.8 lb/100)
80%	162g	(35.6 lb/100)	12 g/d, 3 x/wk	(2.6 lb/100)
-2% from peak	162g	(35.6 lb/100)	8 g/d, 3 x/wk	(1.8 lb/100)
79%	162g	(35.6 lb/100)	None	-
<79%	Reduce		-	None	-

For this flock we are giving a larger challenge feed, because night-time temperature is quite low and there is a problem with maturity related to poorer uniformity. On average this flock may gain a little more weight than example. This will have to be accommodated with a more vigorous post-peak feed withdrawal programme.

Low nutrient dense feed used, with poor ingredient quality control, and so feed composition may be variable. Expected high-low temperatures 28-20 degrees Celsius. Average to good flock uniformity at 20 weeks of age, and past flocks show variable peaks at 80-86%.

Egg production	Base feed	Challenge feed
40%	175g (38.5 lb/100)	None -
65%	175g (38.5 lb/100)	8 g/d, 3 x/wk (1.8 lb/100)
80%	175g (38.5 lb/100)	14 g/d, 3 x/wk (3.0 lb/100)
-2% from peak	175g (38.5 lb/100)	8 g/d, 3 x/wk (1.8 lb/100)
79%	175g (38.5 lb/100)	None -
<79%	Reduce -	None -

For this flock, base feed allowance is increased because a low nutrient dense feed is used, and challenge is fairly aggressive again due to concern over feed quality and poor uniformity.

Challenge feeding can also be used post-peak if there are precipitous declines in egg production related to minor disease challenge or management or environmental stress. Under these conditions, challenges of 10 g/bird/day for two consecutive days are recommended. If no immediate response is seen in egg production, then challenge should be discontinued. If egg production returns to normal, then the challenge should gradually be reduced over the next 2-3 days.

Challenge feeding allows tailoring of feed allocation to suit individual flock needs. Managers should be flexible in actual allocations, although maximum challenge feed allocation needs to coincide with peak egg production. Breeders will respond to a good challenge programme, with sustained peak production and better post-peak persistency. On the other hand, the challenge should not usually represent more than 5% of total feed intake, because excessive challenge will invariably result in obesity and related loss in post-peak performance.

In general, when birds are subjected to such stresses as variable feed quality, mycotoxin challenge and/or fluctuating or extreme environmental temperature, then a high base feed allowance, coupled with aggressive feed challenge, is recommended. On the other hand, lower feed inputs are possible where consistent quality high-energy feeds are used, and where there is good environmental control.

Male Aggression in Broiler Breeders

Dr. Ian Duncan continues to study male aggression in broiler breeders, a problem that appears to be on the increase. His lab initiated a study to look at the effects of genetic strain and feed restriction on male aggression. Two broiler breeder strains and one commercial laying strain of males were tested with a restricted versus an ad-libitum feeding regime. Their data indicated that male aggression was not the result of feeding regime but rather genetic factors apparent in most broiler breeder males

A further study looked at overall aggression of broiler breeder males. Males were compared to a Game strain as well as to a laying strain of males under similar conditions. Aggressiveness towards a male model was also observed. The researchers concluded from their study that aggressiveness of the broiler breeder strain being tested was not due to it being more aggressive since it did not act as aggressively to a male model as did the Game strain males. An additional study looked at female receptiveness to male courtship advances. If females tend to avoid male courtship this could possibly trigger frustration in otherwise normal males. A number of females were tested in a specific maze system where they had a choice between a broiler male and a laying strain male.

It was concluded from the study that females do use male behaviour as a basis for their choice of males and not male appearance. Broiler breeders males were found to be deficient in some elements of courtship behaviour. In practice, this means that females are much less likely to crouch when broiler breeder males approach them and this, in turn, leads to the males chasing the females and forcing themselves on them.

Male aggression in broiler breeder flocks leads to increased bird mortality and also reduced fertility in a flock. Since a significant percentage of the approximately 10% of non-fertile eggs, resulting from the average commercial flock, can be the result of male, female negative interaction, further research into obtaining a 1 to 2% increase in fertile eggs from a flock would be a worthwhile research investment in this type of study.

Pre-Breeder Diets

While most breeding companies provide nutrient specifications for pre-breeder diets, there is considerable variation in their commercial use and application. Using a pre-breeder or pre-lay diet is based on the assumption that the birds nutrient requirements change in this critical period of the birds life. There are certainly major changes occurring in the birds metabolism, hopefully related to ovary and oviduct development, and so this is the basis for a specialised diet at this time. With egg-laying stock, pre-lay diets most often involve a change in calcium nutrition, in order to establish the birds calcium reserves necessary for rapid and sudden onset of eggshell production. The same situation can be applied to heavy breeders today, because with flocks of uniform body weight and with good light management, the subsequent synchronisation of maturity leads to rapid increase in egg numbers up to peak production.

However, most often pre-breeder diets are used in an attempt to "condition" or correct growth and/or body compositional problems that have arisen during the 14-18 week growing period. In these situations managers are perhaps ill-informed of the expectations of merely changing diet specifications at this time.

The Pre-Breeder Period

Although there is no specific pre-breeder "period", most consider the 19-23 week period to be the major transition time for sexual development of the bird. During this time (4 weeks) the pullet is expected to increase in weight by about 570 g. This is somewhat more than the growth expectation of around 340 g for the previous 4 weeks (15-19 week) or growth of around 470 g for the 4 weeks from 23-27 weeks of age.

It is expected that a significant proportion of this growth spurt will be as ovary and oviduct, which are developing in response to light stimulation. A practical complication of this expected development, is that it invariably coincides with move of the pullets from grower to breeder facilities. Under adverse conditions, such as transportation over long distances, heat stress, etc., then birds can lose up to 100 g of body weight at this critical time.

If weight loss is characteristic of such transportation, then pullets should be given an extra feeding. For example, pullets should be moved on an "off-feed" day, but be fed that day in the breeder house after all birds are housed. We cannot allow weight loss at this critical time, and so the question to be answered is - do pre-breeder diets help in this physical move, as well as prime the bird for sexual maturity? Development of the ovary and oviduct require both protein/amino acids and energy (fat) accretion. Nutrients of interest, therefore, are protein and energy, together with increase in calcium for early deposition of medullary bone. However it has never been clearly established that such nutrients need to come from a specially fortified diet versus simply increasing the feed allowance of the grower diet or breeder diet that is introduced prior to maturity.

Calcium Metabolism

With egg-layers, pre-lay diets are used essentially to pre-condition the pullet for impending eggshell production. The very first egg represents a 1.5-2.0g loss in calcium from the body, the source of which is both feed and medullary bone reserve. Today breeder hens are capable of a sustained long clutch length which is necessary to achieve the 85-87% peak production that is now readily attainable. Calcium metabolism is, therefore, very important for the breeder. With Leghorn hens the consequence of inadequate early calcium balance is cage layer fatigue.

Breeders do not show such signs, because they naturally have more exercise, and also have a readily available rich source of diet calcium in the form of their flockmates eggs. Hens have an innate ability to select out calcium, and so improperly fed breeders will eat litter and eggshells in an attempt to balance their diet. However inadequate calcium in the diet does lead to disruption of ovulation, and so these birds stop laying until their meagre calcium reserves are replenished. In a breeder flock, it is the larger bodied, early maturing pullets that are disadvantaged in this manner.

Commercially, we see three different approaches used in pre-breeder calcium nutrition. Firstly, is the use of grower diets that contain just 0.9 - 1.0% calcium being fed up to 5% egg production. This is the system that was used many years ago, and unfortunately is still sometimes used today. At 5% egg production, we do not have 100% of the flock producing at 5% egg production - rather we have closer to 5% of the early maturing pullets producing at close to 100% production.

Pullets can produce just 2-3 eggs with a diet containing 1% calcium. After this time they will eat litter/eggs as previously described, or more commonly simply shut down the ovary. With this approach, birds may in fact be at 10-15% production before the breeder diet is introduced, because no farm system allows for instantaneous change in feed supply as feed tanks are hopefully never completely empty. There is no justification, therefore, for this old system of feed management, because it will be very detrimental to life-time productivity of today's genetic stocks.

The second system involves the classical pre-breeder diet containing around 2% calcium, which is really a compromise situation. It allows for greater medullary bone reserves to develop, without having to resort to the 3.5% calcium as used in a breeder diet.

However 2% calcium is still inadequate for sustained eggshell production - with this diet the breeder can produce 4-6 eggs before ovulation pattern is affected. If a pre-breeder diet is used, therefore, and a moderate calcium level is part of this programme, then the diet must be replaced by the breeder diet before egg production starts. A good rule of thumb is to change from pre-breeder to breeder when the very first egg is noticed, because this occurs usually around 10 days before 1% egg production.

The third option is perhaps the most simple solution, and involves changing from grower to breeder at 1st egg (10 days before 1% production). Having the breeder diet in place before maturity, ensures that even the earliest maturing birds have adequate calcium for sustained early egg production. Proponents of pre-breeder diets suggest that breeder diets introduced early provide too much calcium, and that this contributes to kidney disorders, because the extra ingested calcium must be excreted in the urine.

There is an indication with Leghorns that feeding adult layer diets for 10-12 weeks prior to maturity can adversely affect kidney function, especially if birds are also challenged with infectious bronchitis. However feeding "extra" calcium for one or two weeks prior to maturity has no such effect. It is also interesting to realise that most roosters today are fed high-calcium breeder diets, which provide 4-6x their calcium needs, yet kidney dysfunction is quite rare in these birds.

Body Weight and Size

Body weight and body condition of the bird around the time of maturity, are perhaps the most important criteria that will ultimately influence breeder performance. Body weight and body condition should not really be considered in isolation, although at this time we do not have a good method of readily assessing body condition. Each strain of bird has a characteristic mature body weight that must be reached or surpassed for adequate egg production and egg mass output. In general, prelay diets should not be used in an attempt to manipulate mature body size. The reason for this is that with most flocks it is too late at this stage of rearing to meaningfully influence body weight - all too often prelay diets are used as a crutch for poor rearing management.

However, if birds are underweight when placed in the breeder house, then there is perhaps a need to manipulate body weight prior to maturity. Under controlled environment conditions, this can sometimes be achieved by delaying photostimulation. If prelay diets are then necessarily used in an attempt to correct rearing mismanagement it seems as though the bird is most responsive to energy. This fact likely fits in with the effect of estrogen on fat metabolism, and the significance of fat for liver and ovary development at this time. While such high nutrient density pre-lay diets may be useful in manipulating body weight, it must be remembered that this late growth spurt (if it occurs) will not be accompanied by any meaningful change in skeletal growth.

This means that in extreme cases, where birds are very small in weight and stature at say, 16-18 weeks of age, the end result of using high-nutrient dense prelay diets may well be pullets of correct body weight, but of small stature. These short-shank length pullets seem more prone to prolapse/pick-out, and so this is another example of the limitations in use of classical pre-lay diets.

Use of high-nutrient dense prelay diets to manipulate late growth of broiler breeder pullets does, however, seem somewhat redundant. The reason for this is that with restricted feeding programmes, it is more logical to increase feed allowance than to add the complexity of introducing another diet. The only potential problem of this programme is that in extreme cases feed intake is increased to a level that is in excess of the initial allowance of breeder diet at start of lay ie. ensure that breeders are not subjected to a step-down in feed allocation at time of first egg.

Body Composition

While body composition at maturity may well be as important as body weight at this age, it is obviously a parameter that is difficult to measure. There is little doubt that energy is likely the limiting nutrient for egg production for all classes of birds, and that around peak production, feed may not be the sole source of such energy. Labile fat reserves at this time are, therefore, essential to augment feed sources. These labile fat reserves become critical during situations of heat stress or general hot-weather conditions. Once the bird starts to produce eggs, then its ability to deposit fat reserves is greatly limited. Obviously if labile fat reserves are to be of significance, then they must be deposited prior to maturity.

Egg Weight and Hatchability

It seems as though egg size is ultimately controlled by the size of the yolk that enters the oviduct. In large part this is influenced by body weight of the bird, and so factors described previously for body weight can also be applied to concerns with egg size. There is a general need for as large an early egg size as is possible.

Most attempts at manipulating early egg size have met with limited success. Increased levels of linoleic acid in prelay diets may be of some use, although levels in excess of the regular 1% found in most diets produce only marginal effects on early egg size. From a nutritional standpoint, egg size can best be manipulated with diet protein, and especially methionine concentration. It is logical, therefore, to consider increasing the methionine levels in prelay diets.

For breeders we must also consider egg composition as it relates to early hatchability success. Eggs from young breeders seem to inherently have a hatchability problem, and perhaps this is one of the reasons that we wait for egg size to increase before sending eggs to the hatchery. The reason for this early hatch problem is not fully resolved, but most likely relates in some way to maturity and development of embryonic membranes and their effect on transfer of nutrients from the yolk to the embryo.

However part of this problem may also elate to inadequate transfer of vitamins into the egg. For a number of critical B-vitamins, their concentration in successive eggs does not plateau until after 7-10 eggs have been laid. The effect of pre-lay nutrition on these factors probably warrants further study, but at this time these problems cannot be resolved by simply over-fortifying pre-breeder diets with vitamins or certain fatty acids.

Pre-breeder diets can successfully be used as part of a feeding programme aimed at maximising production potential in young breeders. However any desired increase in nutrient intake prior to maturity can most easily be achieved by simply increasing the feed allowance of either grower or adult breeder diet at this time. If pre-breeder diets are used, then 19-23 weeks seems the most ideal time, assuming 1% production will occur around 24 weeks of age.

Proper Brooder Management

Proper Brooder Management is doing everything possible to promote a comfort zone in the broiler pen which will maximise feed and water intake during the first week of life. Low first week mortality is only possible if we start with quality chicks.

The manager must evaluate the environment by being present in the barn at least three to four times per day. I like to refer to it as working on your M.B.A.: Masters of Business Administration or Management by Being Around.

The formula for success is:

- *Clean Barn:* The barn, including all walls, ceilings, equipment and floors are washed so that all dirt is gone and then the barn is disinfected. This procedure permits the birds to develop antibodies to vaccine which protects the bird for life. Insect and rodent control is an essential component of clean up.

- *Clean Water:* This must be readily available at all times. This means proper height of nipple drinkers, proper pressure of cup drinkers and adequate levels and cleanliness of bell drinkers.

- *Temperature:* Exhaust fans should operate right from day one, but only if there are no drafts at bird level. Ideal temperature at bird level is likely between 87 and 90 degrees Fahrenheit. Chick location dictates comfort and not the thermostat or the computer screen. Chicks are poikilothermic which means that their body temperature is dictated by the environmental temperature. Increasing the barn temperature speeds up growth rate and lowering the temperature slows growth rate. Remember that the relative change in body weight is over 250% within the first 7 days.

- *Feed:* This must be accessible at all times via filled feed pans, paper below feeders, or supplementary feeder trays. Proper texture and quality is imperative. A young flock can have as much as 10 percent of body weight in the crop and gizzard as feed and water within hours of placement.

- *Light:* Light intensity and duration can control weight gain. Metabolic diseases can be significantly reduced if a dramatic light restriction programme is initiated within the first week of life.

Time to Feed Your Breeders

What is the optimum time of day to feed growing and adult hens and roosters? As with most questions, the answer is "it depends" and this situation certainly applies to feeding breeders. For growing birds the variables are ability to observe feeding behaviour, and potential effects of heat stress. With adult birds we have the added factors of eggshell quality and conflict of time associated with mating or nesting.

Growing Birds

For growing birds, feed is eaten in a very short period of time (30 minutes - 2 hrs depending upon age and frequency of feeding) and so choice of feeding time has little real effect on other daily activities. In fact feeding and drinking are the major activities of the immature bird. Most producers will feed growing pullets and roosters early in the morning, especially in warm or hot climates. Digested feed is not utilised with 100% efficiency, and a by-product of such inefficiency is heat production in the birds body.

In most situations this extra heat (sometimes called heat of metabolism, specific dynamic action, or heat increment), peaks about 4-6 hours after feed is eaten. Because of the restricted feeding programme, feeding time is short and predictable, and so the heat of metabolism will consistently peak 4-6 hrs after feeding time. In hot climates peak environmental heat load occurs in the early afternoon, and so there is a distinct disadvantage to having extra heat generated in the birds body at this time.

For this reason we have the common practice of feeding birds at 6-7 am. With such early morning feeding, the heat load of nutrient metabolism occurs before the early afternoon daily high temperature. Alternatively, growing birds could be fed in late afternoon or early evening. However this latter situation does not work well with short-daylengths for growing birds.

With mechanical feeders there is a tendency to feed birds even earlier, sometimes at daylight or when artificial lights are switched on. There are two disadvantages to very early morning feeding. Firstly feeding often occurs before staff are present to observe feeding activity and bird distribution. Under these conditions it is impossible to know if feed is being evenly distributed and if all birds have access to the feed. The second problem, which becomes more acute as birds get older, is the condition of choking, which occurs with a small percentage of older birds, especially on every-day feeding.

This problem can often be resolved by switching on drinkers for at least on hour before feed is available. This is obviously impossible to accomplish if birds are mechanically fed at first daylight or when artificial lights are switched on - pullets seldom drink in the dark period. The ideal feeding time for growing pullets and roosters therefore is early morning, when staff can observe feeding behaviour, and after birds have had access to water for up to 1 hour.

Adult Breeders

Choice of feeding time of adult breeders can influence the production of settable eggs, eggshell quality, fertility and hatch of fertiles. In most instances these factors are a consequence of feeding activity displacing other important daily routines, such as nesting and mating. Breeder hens consume their feed in 2-6 hours each day.

This large variation in feed clean-up time relates to diet energy level, feed texture and perhaps most importantly, environmental temperature. In hot climates breeders often take much longer to eat feed, and this is especially true of high-yield strains. Most managers consider this extended feeding time to be advantageous, because it ensures more even allocation of feed across the flock where even the most timid birds have time to eat.

If breeders are fed early in the morning, then most intense feeding activity will be over by 9 a.m. Again this is ideal in terms of reducing heat load in the early afternoon period. This timing is also ideal in terms of differentiating the main feeding time from nesting activity. Depending upon when lights are switched on in the morning, most eggs are laid in the 9 am - 12 noon period. Feeding at, say 8 am, would, therefore, induce birds to feed at a time when they are usually in the nests. In fact eggs dropped in the area of the feeder are a very good indication of late-morning feeding. Obviously some of these eggs will get broken or become too dirty for setting.

A few years ago there was interest in feeding breeders in the late afternoon. The main advantage is claimed to be an improvement in eggshell thickness, and in fact in many field trials this is found to be true. Improved shell thickness is likely a consequence of the bird eating calcium at a time when shell calcification is starting (for the next days egg) and also the bird having more feed (with calcium) in its crop when lights are switched off. If eggshell quality (thickness) is a problem, then afternoon feeding seems a viable option. Alternatively, birds could be given a "scratch" feed of large particle limestone or oystershell in the late afternoon.

However, late afternoon feeding has a number of potential disadvantages. Firstly there is increase in shell thickness. This should not be a problem as long as incubation setter conditions are adjusted so as to maintain normal moisture loss. In most situations this means reduction in setter humidity to account for less moisture loss through a thicker shell.

A greater concern with later afternoon feeding is potential loss of mating activity, and increase in incidence of body-checked eggs. Mating activity is usually greatest in late afternoon. If hens are more interested in feeding at this time, then there can be reduced mating activity and also more aggression between males. Body-checked eggs are characterised by a distinct band of thickened shell around the middle of the egg (sometimes called belted eggs). This defect is caused by the eggshell breaking during its early manufacture in the birds uterus. The bird repairs the crack, but does so imperfectly. Such eggs have reduced air and moisture transfer characteristics, and usually fail to hatch. The most common cause of body-checked eggs is sudden activity, movement, stress, etc. on the bird. This extra activity takes place when feed is given in late afternoon, and so there will likely be fewer settable eggs produced.

Early morning feeding in breeders is usually recommended because all associated factors and consequences of this practice are positive for the bird and the production of settable eggs. The only concern is with mechanical feeders where there is a temptation to feed too early in the morning, and before staff are present to observe bird activity.

2

Poultry Farming

Poultry farming is the practice of raising domesticated birds such as chickens, turkeys, ducks, and geese, as a subcategory of animal husbandry, for the purpose of farming meat or eggs for food. More than 50 billion chickens are raised annually as a source of food, for both their meat and their eggs. Chickens raised for meat are called broilers, whilst those raised for eggs are called laying hens. In total, the UK alone consumes over 29 million eggs per day. Some hens can produce over 300 eggs a year. Chickens will naturally live for 6 or more years. After 12 months, the hen's productivity will start to decline. This is when most commercial laying hens are slaughtered.

The majority of poultry are raised using intensive farming techniques. According to the Worldwatch Institute, 74 percent of the world's poultry meat, and 68 percent of eggs are produced this way. One alternative to intensive poultry farming is free range farming. Friction between these two main methods has led to long term issues of ethical consumerism. Opponents of intensive farming argue that it harms the environment and creates health risks, as well as abusing the animals themselves. Advocates of intensive farming say that their highly efficient systems save land and food resources due to increased productivity, stating that the animals are looked after in state-of-the-art environmentally controlled facilities. A few countries have banned cage system housing, including Sweden and Switzerland. Consumers can still purchase lower cost eggs from other countries' intensive poultry farms.

Techniques

Free Range

Free range is a term which outside of the United States denotes a method of farming husbandry where the animals are allowed to roam freely instead of being contained in any manner. In the United States, USDA regulations apply only to poultry and indicate that the animal has been allowed access to the outside. The USDA regulations do not specify the quality or size of the outside range nor the duration of time an animal must have access to the outside.

The term is used in two senses that do not overlap completely: as a farmer-centric description of husbandry methods, and as a consumer-centric description of them. Farmers practice free range to achieve free-range or humane certification, to reduce feed costs, to produce a higher-quality product, and as a method of raising multiple crops on the same land. Free range may apply to meat, eggs or dairy farming. In ranching, free-range livestock are permitted to roam without being fenced in, as opposed to fenced-in pastures. In many of the agriculture-based economies, free-range livestock are quite common. There is a diet where the practitioner only eats meat from free-range sources called Ethical Omnivorism which is a type of Semivegetarian.

History

If one allows "free range" to include "herding", free range was a typical husbandry method at least until the development of barbed wire and chicken wire. The generally poor understanding of nutrition and diseases before the twentieth century made it difficult to raise many livestock species without giving them access to a varied diet, and the labour of keeping livestock in confinement and carrying all their feed to them was prohibitive except for high-profit animals such as dairy cattle.

In the case of poultry, free range was the dominant system until the discovery of vitamins A and D in the 1920s, which allowed confinement to be practiced successfully on a commercial scale. Before that, green feed and sunshine (for the vitamin D) were necessary to provide the necessary vitamin content. Some large commercial breeding flocks were reared on pasture into the 1950s. Nutritional science resulted in the increased use of confinement for other livestock species in much the same way.

United States

Free Range Jurisdictions

Traditional American usage equates "free-range" with "unfenced," and with the implication that there was no herdsman keeping them together or managing them in any way. Legally, a free-range jurisdiction allowed livestock (perhaps only of a few named species) to run free, and the owner was not liable for any damage they caused. In such jurisdictions, people who wished to avoid damage by livestock had to fence them out; in others, the owners had to fence them in.

Free Range Poultry

In recent years, with the days of free-range cattle mostly in the past, neither the presence of a "legal fence" surrounding the farm nor the pros and cons of old-time free-range ranching are the main points of interest. Instead, the term "free range" is mainly used as a marketing term rather than a husbandry term, meaning something on the order of, "low stocking density," "pasture-raised," "grass-fed," "old-fashioned," "humanely raised," etc. In poultry-keeping, "free range" is widely confused with yarding, which means keeping poultry in fenced yards. Yarding, as well as floorless portable chicken pens ("chicken tractors") may have some of the benefits of free-range livestock but, in reality, the methods have little in common with the free-range method.

A behavioural definition of free range is perhaps the most useful: "chickens kept with a fence that restricts their movements very little." This has practical implications. For example, according to Jull, "The most effective measure of preventing cannibalism seems to be to give the birds good grass range." De-beaking was invented to prevent cannibalism for birds not on free range, and the need for de-beaking can be seen as a litmus test for whether the chickens' environment is sufficiently "free-range-like."

De-beaking does not address the fact that cannibalistic tendencies stem from stress hormone elevation, which results directly from overcrowding conditions, and that these stress hormones inhibit the conditions necessary for the development of omega-3 fatty acids and drastically diminish the nutritive value of both the meat and eggs. As a result of that, the addition of omegas to chicken feed has been an attempt to address the inability of chickens to have enough access to insects and seeds during daily forage. Chicken's peck at each other out of aggression to disperse their population out to more naturally sustainable levels within a given environment. A rule of thumb is that if debeaking is required to address chickens' excessive pecking and cannibalism behaviour, the chickens are under stress and the meat and egg products, as a result, are of lesser quality.

The U.S. Department of Agriculture Food Safety and Inspection Service (FSIS) requires that chickens raised for their meat have access to the outside in order to receive the free-range certification. There is no requirement for access to pasture, and there may be access to only dirt or gravel. Free-range chicken eggs, however, have no legal definition in the United States. Likewise, free-range egg producers have no common standard on what the term means.

The USDA has no specific definition for "free-range" beef, pork, and other non-poultry products. All USDA definitions of "free-range" refer specifically to poultry. No other criteria-such as the size of the range or the amount of space given to each animal-are required before beef, lamb, and pork can be called "free-range". Claims and labelling using "free range" are therefore unregulated. The USDA relies "upon producer testimonials to support the accuracy of these claims."

In a December 30, 2002 Federal Register notice and request for comments (67 Fed. Reg. 79552), USDA's Agricultural Marketing Service proposed "minimum requirements for livestock and meat industry production/marketing claims". Many industry claim categories are included in the notice, including breed claims, antibiotic claims, and grain fed claims. "Free Range, Free Roaming, or Pasture Raised" would be defined as "livestock that have had continuous and unconfined access to pasture throughout their life cycle" with an exception for swine ("continuous access to pasture for at least 80% of their production cycle"). This proposed rule making is still in play.

In a May 12, 2006 Federal Register notice (71 Fed. Reg. 27662), the agency presented a summary and its responses to comments received in the 2002 notice, but only for the category "grass (forage) fed" which the agency stated was to be a category separate from "free range." Comments received for other categories, including "free range," are to be published in future Federal Register editions.

The broadness of "free range" in the U.S. has caused some people to look for alternative terms. "Pastured poultry" is a term promoted by farmer/author Joel Salatin for broiler chickens raised on grass pasture for all of their lives except for the initial brooding period. The Pastured Poultry concept is promoted by the American Pastured Poultry Producers' Association (APPPA), an organisation of farmers raising their poultry using Salatin's principles.

Alternative terminology can also be used to make high-density confinement sound more palatable. For example: *cage-free, free-*

running, free-roaming, naturally nested, etc. are used as an alternative to the technical term, *high-density floor confinement.* Whether high-density floor confinement is more humane than high-density cage confinement is arguable, but in any event, high-density confinement (of whatever type) is the antithesis of free range.

European Union

The European Union regulates marketing standards for egg farming which specifies the following (cumulative) minimum conditions for the free-range method:

- hens have continuous daytime access to open-air runs, except in the case of temporary restrictions imposed by veterinary authorities,

- the open-air runs to which hens have access is mainly covered with vegetation and not used for other purposes except for orchards, woodland and livestock grasing if the latter is authorised by the competent authorities,

- the open-air runs must at least satisfy the conditions specified in Article 4(1)(3)(b)(ii) of Directive 1999/74/EC whereby the maximum stocking density is not greater than 2500 hens per hectare of ground available to the hens or one hen per $4m^2$ at all times and the runs are not extending beyond a radius of 150 m from the nearest pophole of the building; an extension of up to 350 m from the nearest pophole of the building is permissible provided that a sufficient number of shelters and drinking troughs within the meaning of that provision are evenly distributed throughout the whole open-air run with at least four shelters per hectare.

Otherwise, egg farming in EU is classified into 4 categories: Organic (ecological), Free Range, Barn, and Cages.) The mandatory labelling on the egg shells attributes a number (which is the first digit on the label) to each of these categories: 0 for Organic, 1 for Free Range, 2 for Barn and 3 for Cages.

There are EU regulations about what free-range means for laying hens and broilers (meat chickens) as indicated above. However, there are no EU regulations for free-range pork, so pigs could be indoors for some of their lives. In order to be classified as free-range, animals must have access to the outdoors for at least part of their lives.

United Kingdom

Free-range pregnant sows are kept in groups and are often provided with straw for bedding, rooting and chewing. Around 40% of UK sows are kept free-range outdoors and farrow in huts on their range. Egg laying hens Cage-free egg production includes barn, free-range and organic systems. In the UK, free-range systems are the most popular of the non-cage alternatives, accounting for around 28% of all eggs, compared to 4% in barns and 6% organic. In free-range systems, hens are housed to a similar standard as the barn or aviary.

Turkeys Free-range turkeys have continuous access to an outdoor range during the daytime. The range should be largely covered in vegetation and allow more space. Access to fresh air and daylight means better eye and respiratory health. The turkeys are able to exercise and exhibit natural behaviour resulting in stronger, healthier legs. Free-range systems often use slower-growing breeds of turkey.

Free Range Rearing of Chickens Free Range Rearing of birds is now being pioneered in the UK by various poultry rearing farms allowing the birds outside from just a few weeks of age and not just after the birds have been reared in barns and allowed out at 16 weeks. Allowing the birds outside space from just 4 weeks old.

Yarding

In poultry keeping, yarding is the practice of providing the poultry with a fenced yard in addition to a poultry house. Movable yarding is a form of managed intensive grasing. Yarding is often confused with free range. The distinction is that free-range poultry are either totally unfenced, or the fence is so distant that it has little influence on their freedom of movement.

Historical Practice

Before the discovery of vitamins A and D in the 1920s, green feed and sunshine were essential to the health of poultry. Vitamin D was synthesised from sunlight on the skin (as with humans), while Vitamin A was obtained through green forage plants such as grass. Yards small enough to be fenced economically were soon stripped of palatable green forage and become barren. This is followed by a build-up of manure, parasites, and other pathogens.

Free range husbandry was the most common method in these early days. Most farms had only a small free-range barnyard flock.

Larger flocks were kept in small houses build on skids, which were dragged periodically to a fresh piece of ground. This method is similar to the modern practice of pastured poultry.

Experts of the day estimated the sustainable level to be about fifty hens per acre (80 m² per hen), with one hundred hens per acre (40 m² per hen) as an absolute upper limit if special care was taken. These levels are sustainable in the sense that the turf can make use of the nutrients in the manure left behind by the chickens, and in the sense that, at this stocking density, the chickens will not completely destroy the turf through scratching.

At the Oregon Station on clay soil it was found that the day droppings from 200 laying hens on an acre (20 m² per hen) in four years made the soil too rich for the successful growth of cereal crops where cropping the ground was done every other year. The night droppings were put on other land. If the soil contains too much manure for the crops it is safe to assume that it is not in the best condition for poultry. Sooner or later it is bound to show not only in a failure of grain crops but in failure of poultry crops. For a permanent system under average conditions of soil and climate the following points are suggested for consideration.

1. Maximum number of fowls per acre: 100 laying hens (40 m² per hen).

2. Disposing of the night droppings on other land.

3. Dividing the ground into at least two divisions or yards, and growing a crop on each yard at least every other year. In sections where crops may be grown every years the number of fowls may be increased.

4. Growing crops that will use up the maximum amount of manure.

5. Keeping the ground vacant (of chickens) at least six months in the year.

6. Thorough underdrainage, where necessary, to carry off surplus water.

It is not assumed assumed that as many as 500 hens may not be profitably kept on an acre (8 m² per hen) for a few years under favourable conditions. It has been done, but it is a different matter when it is planned to make a permanent business of it.

Because fifty hens per acre represents 800 square feet (74 m²) per hen (80 m² per hen), while the density inside the house at the time was normally four square feet per hen (0.4 m² per hen), this required that the yard be 200 times wider than the house, assuming a yard on one side of the house. That is, a house 20 feet (6 m) wide required a yard 400 feet (120 m) wide to provide the necessary area. This would normally be provided as two yards, one on either side of the house, each 200 feet (60 m) wide. In reality, such yards are expensive to fence, and the chickens spend most of their time on the portion closest to the house, so sustainability was never achieved in practice except with portable houses, which were moved periodically to fresh ground. Yarded operations were operated with unsustainably small yards that were quickly denuded and which received excessive levels of manure.

The use of multiple yards, frequent plowing, and liberal use of lime would allow higher stocking levels to be used, since plowing and liming would allow much of the nitrogen to escape from the soil.

The following is typical advice for the successful use of yards in the Thirties and Forties:

> *All poultrymen should realise that there are no known*
> *substitutes for sunshine and young green grass in keeping*
> *poultry in the best possible state of health and in*
> *promoting growth and maintaining egg production.*
> *Where sunshine and green grass cannot be provided, as*
> *in the case of birds kept in strict confinement, the best*
> *possible substitutes must be provided. In the case of*
> *most farm and many commercial flocks, however, the*
> *growing stock is reared on range, and the adult birds*
> *are given yards or allowed to roam at will. If the*
> *staggering losses among growing chicks and laying birds*
> *that occur annually are to be reduced materially, better*
> *methods of flock management must be employed.*

The losses from mortality are due largely to internal parasites and diseases of one kind or another. Bare ground over which the chickens have run for some time, mud puddles, and stagnant water are the chief sources of the spread of diseases, most of which are filth borne.... The mortality that usually occurs in growing and adult stock may be materially reduced by providing the birds with an alternate yarding system. Probably the best arrangement is to provide each colony brooder and each laying house with three yards (3 m) which the birds would be allowed to use every 3 or 4 weeks. By alternating the birds in the yards every 3 or 4 weeks each yard is kept reasonably sanitary, especially if the soil in the immediate vicinity of the house is cultivated and treated with lime, and young green grass is available for the birds throughout the season. The importance of clean range for both birds and adult stock cannot be emphasised too strongly... For adult stock a good grass sward can be maintained on fertile soil, allowing about 200 birds to the acre (20 m² per hen).

Nutritional advances increasingly turned yarding into a liability, and it fell out of favour. Free range continued to be used, especially for breeding flocks and for pullets before they reached laying age, because of the lower rate of disease and greater overall health of grass-reared chickens.

Breeding flocks (which lay eggs destined for incubation) are always given a better diet than flocks laying table eggs, since a diet that will produce table eggs cheaply will not provide eggs that hatch well. For some time after confined laying flocks produced table eggs satisfactorily, breeding flocks benefited from free range

In Britain, Geoffrey Sykes developed a new yarding system in the Fifties. This used a small yard covered with a thick layer of straw, with more straw added frequently. He also recommended that shade and a windbreak be provided by a solid fence around the yard, or by other means, such as rows of haybales. Once a year, the old straw was removed by a front-end loader or similar machinery. This method eliminated mud and pathogens. It was later forgotten because the industry moved to high-density confinement before the method was widely established.

Recent Practice

Today, commercial poultry producers generally call yarding free range on their labels. This conflation of two very different techniques has led to confusion. The vast majority of "free-range" operations are really yarded.

Pastured poultry, as promoted by the APPPA, the American Pastured Poultry Producers Association, and author/farmer Joel Salatin, takes a different approach, attempting to achieve the benefits of free range while using penning or yarding. The key element of Pastured poultry is the use of portable housing and the optional use of portable electric fencing. By moving the house and yard frequently, perhaps daily, all the disadvantages of permanent yards are eliminated.

Intensive Chicken Farming

Figure 1: Egg-laying chickens in battery cages

Figure 2: Egg-laying chicken 5 days out of battery cage

In egg-producing farms, birds are typically housed in rows of battery cages. Environmental conditions are automatically controlled, including light duration, which mimics summer daylength. This stimulates the birds to continue to lay eggs all year round. Normally, significant egg production only occurs in the warmer months. Critics argue that year-round egg production stresses the birds more than normal seasonal production.

Figure 3: Broilers in a production house

Meat chickens, commonly called broilers, are floor-raised on litter such as wood shavings or rice hulls, indoors in climate-controlled housing. Poultry producers routinely use nationally approved medications, such as antibiotics, in feed or drinking water, to treat disease or to prevent disease outbreaks arising from overcrowded or unsanitary conditions. In the U.S., the national organisation overseeing chicken production is the Food and Drug Administration (F.D.A.). Some F.D.A.-approved medications are also approved for improved feed utilisation.

In egg-producing farms, cages allow for more birds per unit area, and this allows for greater productivity and lower space and food costs, with more efforts put into egg-laying. In the U.S., for example, the current recommendation by the United Egg Producers is 67 to 86 in^2 (430 to 560 cm^2) per bird, which is about 9 inches by 9 inches. Modern poultry farming is very efficient and allows meat and eggs to be available to the consumer in all seasons at a lower cost than free range production, and the poultry have no exposure to predators.

The cage environment of egg producing does not permit birds to roam. The closeness of chickens to one another frequently causes cannibalism. Cannibalism is controlled by de-beaking (removing a portion of the bird's beak with a hot blade so the bird cannot effectively peck). However, de-beaking does not fully prevent cannibalism it just reduces the damage.

Most battery chickens are missing 30-70% of plumage by the time that they are spent. Another condition that can occur in prolific egg laying breeds is osteoporosis. This is caused from year-round rather than seasonal egg production, and results in chickens whose legs cannot support them and so can no longer walk. During egg production, large amounts of calcium are transferred from bones to create eggshell. Although dietary calcium levels are adequate, absorption of dietary calcium is not always sufficient, given the intensity of production, to fully replenish bone calcium.

Under intensive farming methods, a meat chicken will live less than six weeks before slaughter. This is half the time it would take traditionally. This compares with free-range chickens which will usually be slaughtered at 8 weeks, and organic ones at around 12 weeks.

In intensive broiler sheds, the air can become highly polluted with ammonia from the droppings. This can damage the chickens' eyes and respiratory systems and can cause painful burns on their legs (called hock burns) and feet. Chickens bred for fast growth have a high rate of leg deformities because they cannot support their increased body weight. Because they cannot move easily, the chickens are not able to adjust their environment to avoid heat, cold or dirt as they would in natural conditions. The added weight and overcrowding also puts a strain on their hearts and lungs. In the U.K., up to 19 million chickens die in their sheds from heart failure each year.

Indoor with Higher Welfare

Chickens are kept indoors but with more space (around 12 to 14 birds per square metre). They have a richer environment for example with natural light or straw bales that encourage foraging and perching. The chickens grow more slowly and live for up to two weeks longer than intensively farmed birds. The benefits of higher welfare indoor systems are the reduced growth rate, less crowding and more opportunities for natural behaviour.

Issues with Poultry Farming

Humane Treatment

Figure 4: Chickens transported in a truck.

Animal welfare groups have frequently criticised the poultry industry for engaging in practices which they believe to be inhumane. Many animal rights advocates object to killing chickens for food, the "factory farm conditions" under which they are raised, methods of transport, and slaughter. Compassion Over Killing and other groups have repeatedly conducted undercover investigations at chicken farms and slaughterhouses which they allege confirm their claims of cruelty.

Conditions in intensive chicken farms may be unsanitary, allowing the proliferation of diseases such as salmonella and E. coli. Chickens may be raised in total darkness; hens are most often kept in crowded wire battery cages with space less than that of a sheet of paper per hen, as opposed to cage-free or free range. Rough handling and crowded transport during various weather conditions and the failure of existing stunning systems to render the birds unconscious before slaughter have also been cited as welfare concerns.

Another animal welfare concern is the use of selective breeding to create heavy, large-breasted birds, which can lead to crippling leg disorders and heart failure for some of the birds. Concerns have been raised that companies growing single varieties of birds for eggs or meat are increasing their susceptibility to disease.

A common practice among hatcheries is the culling of newly born male chicks of egg laying breeds, since they don't lay eggs, and do not grow fast enough to be profitable for meat.

Debeaking

Laying hens are routinely de-beaked when young to prevent fighting and feather pecking. Animal rights activist claim this is bad because beaks are sensitive, and the usual practice of trimming them without anaesthesia is considered inhumane by some. De-beaked chickens will peck much less than chickens with beaks, which animal behaviourist Temple Grandin attributes to guarding against pain. The chicken industry says that de-beaking is not painful. Others argue that the procedure causes life-long chronic pain and discomfort and decreased ability to eat or drink.

Intelligence

Some groups which advocate for more humane treatment of chickens claim that chickens are intelligent. Dr. Chris Evans of Macquarie University claims that their range of 20 calls, problem solving skills, use of representational signaling, and the ability to recognise each other by facial features demonstrate the intelligence of chickens.

Antibiotics

Antibiotics have been used on poultry in large quantities since the 1940s, when it was found that the by-products of antibiotic production, fed because the antibiotic-producing mold had a high level of vitamin B_{12} after the antibiotics were removed, produced higher growth than could be accounted for by the vitamin B_{12} alone. Eventually it was discovered that the trace amounts of antibiotics remaining in the by-products accounted for this growth.

The mechanism is apparently the adjustment of intestinal flora, favouring "good" bacteria while suppressing "bad" bacteria, and thus the goal of antibiotics as a growth promoter is the same as for probiotics. Because the antibiotics used are not absorbed by the gut, they do not put antibiotics into the meat or eggs.

Antibiotics are used routinely in poultry for this reason, and also to prevent and treat disease. Many contend that this puts humans at risk as bacterial strains develop stronger and stronger resistances. Critics point out that, after six decades of heavy agricultural use of antibiotics, opponents of antibiotics must still make arguments about theoretical risks, since actual examples are hard to come by. Those antibiotic-resistant strains of human diseases whose origin is known originated in hospitals rather than farms.

A proposed bill in the United States Congress would make the use of antibiotics in animal feed legal only for therapeutic (rather than preventative) use, but it has not been passed. However, this may present the risk of slaughtered chickens harbouring pathogenic bacteria and passing them on to humans that consume them.

In October 2000, the U.S. Food and Drug Administration (FDA) discovered that two antibiotics were no longer effective in treating diseases found in factory-farmed chickens; one antibiotic was swiftly pulled from the market, but the other, Baytril, was not. Bayer, the company which produced it, contested the claim and as a result, Baytril remained in use until July 2005.

To prevent any residues of antibiotics in chicken meat, any given antibiotics are required to have a "withdrawal" period before they can be slaughtered. Samples of poultry at slaughter are randomly tested by the FSIS, and shows a very low percentage of residue violations.

Arsenic

Chicken feed can also include Roxarsone, an antimicrobial drug that also promotes growth. Roxarsone was used as a broiler starter by about 70% of the broiler growers between 1995 to 2000. The drug has generated controversy because it contains arsenic, which is highly toxic to humans. This arsenic could be transmitted through run-off from the poultry yards. A 2004 study by the U.S. magazine Consumer Reports reported "no detectable arsenic in our samples of muscle" but found "A few of our chicken-liver samples has an amount that according to EPA standards could cause neurological problems in a child who ate 2 ounces of cooked liver per week or in an adult who ate 5.5 ounces per week." The U.S. Food and Drug Administration (FDA), however, is the organisation responsible for the regulation of foods in America, and all samples tested were "far less than the... amount allowed in a food product."

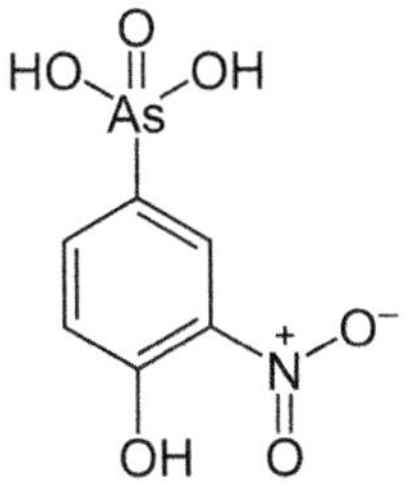

Figure 5: Roxarsone, a controversial arsenic compound used as a nutritional supplement for chickens.

Growth Hormones

Hormone use in poultry production is illegal in the United States. Similarly, no chicken meat for sale in Australia is fed hormones. Several scientific studies have documented the fact that chickens grow rapidly because they are bred to do so, not because of growth hormones. A small producer of natural and organic chickens confirmed this assumption:

> "If this were 1948, you might have something to worry about. Using hormones to boost egg production was a brief fad in the Forties, but was abandoned because it didn't work. Using hormones to produce soft-meated roasters was used to some extent in the Forties and Fifties, but the increased growth rates of broilers made the practice irrelevant—the broilers got as big as anyone wanted them to get when they were still young enough to be soft-meated without chemicals.

The only hormone that was ever used in any quantity on poultry (DES) was banned in 1959, after everyone but a few die-hard farmers had given them up as a silly idea. Hormones are now illegal in poultry and eggs. The people who advertise "No hormones" are either woefully ignorant or are indulging in cynical fear-mongering, maybe both."

E. Coli

According to Consumer Reports, "1.1 million or more Americans are sickened each year by undercooked, tainted chicken." A USDA study discovered *E. coli* in 99% of supermarket chicken, the result of chicken butchering not being a sterile process. However, the same study also cautions that the type of *E. coli* turned up was in every case a non-lethal form distinct from the more dangerous "O157:H7" strain. Many of these chickens, furthermore, had relatively low levels of contamination.

Feces tend to leak from the carcass until the evisceration stage, and the evisceration stage itself gives an opportunity for the interior of the carcass to receive intestinal bacteria. (So does the skin of the carcass, but the skin presents a better barrier to bacteria and reaches higher temperatures during cooking). Before 1950, this was contained largely by not eviscerating the carcass at the time of butchering, deferring this until the time of retail sale or in the home.

This gave the intestinal bacteria less opportunity to colonise the edible meat. The development of the "ready-to-cook broiler" in the 1950s added convenience while introducing risk, under the assumption that end-to-end refrigeration and thorough cooking would provide adequate protection. *E. coli* can be killed by proper cooking times, but there is still some risk associated with it, and its near-ubiquity in commercially farmed chicken is troubling to some. Irradiation has been proposed as a means of sterilising chicken meat after butchering.

Avian Influenza

There is also a risk that crowded conditions in chicken farms will allow avian influenza (bird flu) to spread quickly. A United Nations press release states: "Governments, local authorities and international agencies need to take a greatly increased role in combating the role of factory-farming, commerce in live poultry, and wildlife markets which provide ideal conditions for the virus to spread and mutate into a more dangerous form..."

Efficiency

Farming of chickens on an industrial scale relies largely on high protein feeds derived from soybeans; in the European Union the soybean dominates the protein supply for animal feed, and the poultry industry is the largest consumer of such feed. Two kilograms of grain must be fed to poultry to produce 1 kg of weight gain. However, for every gram of protein consumed, chickens yield only 0.33 g of edible protein.

Economic Factors

Changes in commodity prices for poultry feed have a direct effect on the cost of doing business in the poultry industry. For instance, a significant rise in the price of corn in the United States can put significant economic pressure on large industrial chicken farming operations.

World Chicken Population

The Food and Agriculture Organisation of the United Nations estimated that in 2002 there were nearly sixteen billion chickens in the world, counting a total population of 15,853,900,000. The figures from the *Global Livestock Production and Health Atlas* for 2004 were as follows:

1. China (3,860,000,000)
2. United States (1,970,000,000)
3. Indonesia (1,200,000,000)
4. Brazil (1,100,000,000)
5. Mexico (540,000,000)
6. India (495,000,000)
7. Russia (340,000,000)
8. Japan (286,000,000)
9. Iran (280,000,000)
10. Turkey (250,000,000)
11. Bangladesh (172,630,000)
12. Nigeria (143,500,000)

Identifying Broiler Breeder Management-Nutrition Interactions To Optimise Chick Production

Broiler breeder hens grow more efficiently and are leaner than ever before due to positive results from broiler genetic selection strategies. Effective ovary management is an integral part of a successful breeder management programme. Optimum ovarian morphology and reproductive efficiency is realised when the pre-maturational period is managed very conservatively in terms of increases in feed allocation. Feed allocation programmes need to be based on a solid understanding of the reproductive physiology of these birds and defining factors influencing their growth and reproductive efficiency. By identifying 'reproductive attitudes' of individuals and their incidence in a population, more effective refinement of broiler breeder management strategies will be possible.

Broiler breeders have a lot expected of them. This parent must have the genetics for rapid and efficient growth, and yet exhibit a high rate of egg production to supply the next generation of broiler chicks. While breeding programmes have resulted in annual improvements in broiler growth, breast muscle yield, feed efficiency and disease resistance decades of selection for meat production traits have impaired the reproductive ability of broiler parents.

By the 1970's there were indications that growth selection negatively affected egg production traits. In their comparison of high and low juvenile body weight lines, Udale et al. (1972) showed that the high weight lines had increased rates of internal ovulation and defective egg production (36% versus 2% in high and low weight birds, respectively).

Broiler breeders are feed restricted from early in life to Optimise reproductive performance. These birds have been demonstrated to be prone to multiple hierarchies - a situation that could be alleviated by feed restriction (Van Middelkoop, 1971). The use of feed restriction in modern broiler breeders has limited the expression of such negative reproductive responses to cases where birds have been overfed at a time when the ovary is especially sensitive to excess nutrient intake.

These birds are changing in terms of their reaction to lighting programmes as well (Joseph et al., 2002). Whereas specialised genetic selection has meant that egg production is not remarkably different from what it was a few years ago, hatching egg producers have had to work hard at fine tuning strain specific procedures for nutrient allocation and photoperiod management. As the modern broiler breeder continues to change due to the impact of genetic selection for improved growth efficiency and meat yield, there is value in understanding how our management priorities have changed along with the bird.

Managing Ovarian Follicular Dynamics

The ovary is the core of a successful broiler breeder. This is the point where the internal balance between growth and reproduction will interface with external management methods. In general, a hen with a well coordinated reproductive system will have an ideal state of physical maturity and number of large ovarian follicles (ovarian, yolky follicles greater than 10 mm diameter) at sexual maturity to support a strong, sustainable reproductive effort.

However, many external factors can affect egg production. Specific feed ingredients, bird age, and flock management decisions can directly affect semen quality, the oviduct environment, and the egg environment. Furthermore, even small degrees of over or under feeding have been shown to negatively impact egg and chick production. Something as simple as not adjusting feed allocation for temperature changes can affect nutrients available for reproduction and storage due to altered metabolic requirements. Understanding the ovarian function of the chicken and its interaction with nutritional status, age, and strain is essential to the effective production of fertile eggs with a high probability of hatching.

Both bird age and feeding level can influence how the ovary develops. Extra feed can highly stimulate large yellow follicle development, although this stimulation is greater at younger photostimulation ages. Broiler breeders are thought to be most responsive (or 'estrogenic') in the weeks immediately following photostimulation. During this period, the estrogen output by the ovarian follicles is increasing, and will not decrease to its mature baseline concentration until egg production is underway (Bacon et al., 1980). Allowing excess nutrients at this time may potentially be the most detrimental for normal ovary development. The most critical period for avoiding sudden increases or excessive feed allocation appears to be 2 to 4 wk after photostimulation. This period is a time of flux in the management of nutrients, as the bird switches from primarily growth to a reproductive state. The reproductive and metabolic hormone pathways do not appear to be mature enough to withstand the challenge of a sudden increase in nutrient intake.

The ad libitum feeding of broiler breeder females from photostimulation results in an 80 to 100% greater increase in Luteinising Hormone (LH) and Follicle Stimulating Hormone (FSH) than in feed restricted birds by 3-d after photostimulation resulting in an accelerated sexual maturation process (Renema et al., 1999). This feed-driven accelerated sexual maturation process is typically also associated with elevated ovarian large yellow follicle numbers. The primary influence on how many large, yolky follicles form on the ovary is body weight. However, when you compare birds of the same size, the one consuming more feed will have more large follicles (Hocking, 1993).

A flock with poor ovarian control will have a normal peak production, but undergo more rapid decline in egg production. Unfortunately this may be indicative of either too few or too many large yellow follicles. Having too few large follicles can result in gaps in laying sequences and hence shorter than normal sequences.

This is a problem seen in hens undergoing follicular atresia (follicle dissolution), photostimulated too early, or being underfed. Having too many large ovarian follicles is a problem associated with obesity or overfeeding (Yu et al., 1992). The hen can grow even more overweight as the rate of egg production remains low or goes into early decline due to excess feed intake.

Bird Variability and Reproductive Efficiency

If consumers could dictate how the poultry industry functioned, variation in growth, conformation and efficiency would not exist. The food industry is striving to achieve an increased level of product consistency and bird-to-bird variation is not wanted. However, this same variability is essential for the continued movement in genetic selection programmes towards more desirable bird-types. Without natural variation, genetic progress would be stalled at the current bird-types. The 'ideal hen' with the perfect balance between growth and egg production traits, is more of an exception to the rule rather than the common occurrence in the broiler breeder barn.

Commercially, body weight, egg production, fertility and hatchability values are collected on a barn or farm basis. However, variation among individual hens affects how they do or do not respond to environmental or treatment conditions. For example, a high proportion of the unsettable eggs produced by a flock are from a small number of birds. Individual response to environmental cues can also vary. Proudman and Siopes (2002) reported no change, moderate sensitivity, or extreme sensitivity in turkey hens given a short-term reduction in light period. A range of variability in egg weight exists within individual hens. Egg weight-based differences in hatchability have been reported to be more closely related to deviations in egg weight from the individual hen mean egg weight than from the population mean egg weight (Wilson, 1991). University of Alberta egg weight data shows a 15 g range in breeder flock egg weight at a given age, while variation within individual hens ranges from 5 to 15 g (Renema, unpublished observations).

There are good reasons to work towards achieving a uniform flock - particularly at the end of the pullet phase, when a high proportion of the birds will ideally respond similarly to the photostimulatory cue. Whereas delaying photostimulation to allow the smaller pullets more time to become physically mature may appear counterproductive for the productivity of the larger birds, most hens will compensate with a greater rate of lay that will typically result in similar overall production.

Early Pullet Growth Affects Body Weight Uniformity

Early feed management practices are believed to have a long-term impact on frame size, fleshing, and body weight uniformity. Falling short of 4 week weight or protein intake targets is believed to adversely affect frame size and hen weight management. To test the impact of early feeding on frame size and fleshing, a study was performed comparing the effects of full feeding broiler breeder pullets until 1 or 3 weeks of age on frame size, fatness and fleshing at 4, 8, 12, and 16 weeks of age.

The transition to feed restriction was smoothed by starting it already at 1 week of age. Ross 308 pullets (720) were placed at day of hatch (8 pens) and provided ad libitum access to feed for 1 (1WK) or 3 weeks (3WK) of age. Body weight was recorded twice/week to allow the growth profiles to be gradually converged (target of 8-10 weeks of age). At 4, 8, 12, and 16 weeks, external carcass and fleshing scores were recorded for all birds, and 14 birds/pens were dissected for assessment of muscle mass, fatness, and reproductive development.

By 3 weeks of age, the daily gain of the 3WK pullets was double that of the 1WK pullets, which resulted in significantly more weight and fleshing at 4 weeks of age. The 3WK birds weighed 30% more, had a larger shank and keel length, and carried a higher proportion of breast muscle (12.6% compared to 11.5%) than the 1WK birds. The groups still differed in some traits at 8 weeks of age, but were similar in comparisons after 10 weeks of age, when the body weight profiles met. The body weight uniformity of the 1WK birds was better than that of the 3WK birds from 14 weeks of age (CV of 13.0% compared to 16.7%). At 16 weeks of age, the frame size and fleshing of the birds were indistinguishable, and body weight uniformity was worse in the 3WK birds. In sister pullets from this study that continued on to sexual maturity, there were no differences in how these birds entered lay, or in carcass or reproductive traits at that time (Renema, unpublished observations). Maintaining very tight control body weight profiles through more frequent feed allocation decisions may contribute more to the effective management of broiler breeder productivity than simply following feeding guides.

Influence of Growth History on Feed Utilisation

In many ways, the post-peak feeding period is not as critical as the pullet phase, or the sexual maturation and early production period. The condition of the reproductive system at the onset of production has long-term effects on the potential reproduction of the hen. While the post-peak period (from approximately 32 to 60 weeks of age) is the most financially important, production problems at this time are generally the result of damage done earlier in the life of the breeder. For example, overfeeding of hens for as little as 2 weeks between 23 and 31 weeks of age has been found to reduce fertility and hatchability throughout production (Ingram and Wilson, 1987).

In a recent study, we grew broiler breeders on rapid or slow BW profiles between 5 and 12 weeks of age, and merged the BW targets by 32 weeks of age. By the time birds were dissected at 58 weeks of age there was a clear, long-term effect of pre-peak nutrient allocation on muscle, fat and the ovary. The hens fed the early aggressive, HIGH profile carried a slightly increased proportion of breast muscle than those of most of the less aggressive treatments by sexual maturity (data not presented).

At 58 weeks of age, it was striking how the HIGH profile hens still had more breast muscle, but also had a smaller ovary and less abdominal fat. These hens entering lay with extra fleshing maintained this extra muscle mass at the expense of the ovary. Birds of all treatments were maintaining very little fat. With reduced nutrient stores to rely on to cover shortfalls in nutrient requirements and the easy diversion of nutrients into fleshing, this shows how effective management continues to demand a higher degree of attention to detail.

It can be difficult to formulate diets to Optimise egg production, fertility, and hatchability as little is known about the nutritional requirements of the embryo. Growthselected stocks have low immuno-responsiveness (Siegel et al., 1984) due to either inadvertent negative selection pressure combined with growth efficiency selection. The developing embryo is especially sensitive to vitamin deficiency, which will result in death, malformation or some other atypical response. The importance of the macro-minerals and electrolytes for the maintenance of hen productivity is well established. This has meant that study of the carry-over of minerals from hen to chick for the enhancement of early growth and immunity is being done with the trace minerals (Kidd, 2003). Manganese, selenium, zinc, and vitamin E in the maternal diet have been identifies as important for the improved immunity of the progeny and ability of the embryo to survive incubation.

Linking Growth and Reproductive Efficiency

Examination of individual growth and egg production profiles can reveal breeder hens that do not fit the classic balance between birds that lay well at the expense of growth, or grow very well while producing fewer eggs. Following a recent study, an individual 'snapshot' of production was compiled using feed intake, actual and target BW profiles, BW gains and losses, egg production, egg weight, and final carcass fat and protein concentrations.

Each hen appears to have a different balance between the pull to lay eggs or to grow. However, some hens were present that were able to lay eggs very well and gain body weight relative to the flock average. Conversely, some birds on the same feed allocation grew very poorly and did not produce many eggs. To create an objective score of these various 'reproductive attitudes' scoring the balance between the pull to lay eggs or to grow, hens were also scored for overall efficiency.

Variation in feed utilisation among hens exists that cannot be explained by metabolic body weight, body weight gain, and egg mass output. This variation can be reflected in hen Residual Feed Intake (RFI). The RFI is the difference between observed and predicted feed intake and is a measure of feed efficiency that estimates the remaining part of the variation in feed consumption that cannot be accounted for by changes in growth, maintenance or egg production.

Birds that consumed less feed than we calculated they needed had a negative residual feed intake. These are great birds because they are more efficient than we calculate they should be. Birds that consume more feed than we calculate they need for their activities end up with a positive residual feed intake - meaning that they are consuming more feed than we calculate they should. This remaining variation can be the result of differences in: 1) maintenance requirements, 2) partial efficiency in energy utilisation, 3) energy demanding processes not accounted for, and 4) measuring errors.

If the reproductive attitude of the hen demonstrates good overall efficiency, will this carry forward to the broiler offspring? Defining the relationship between maternal efficiency and the quality of the broiler offspring is an essential step in providing support to future decisions on future breeder management methods and the provision of high quality broilers with desirable meat and growth traits.

Causes of Variation in Efficiency

One of the major sources of variation of RFI in laying hens is a difference in maintenance requirements among birds. Gabarrou et al. (1998) reported that a great part of variation in maintenance requirements in laying hens may be due to feeding activity, with less efficient hens demonstrating a higher regulatory thermogenesis resulting in dissipation of excess energy as heat. The liver, gut, and reproductive tract of broiler breeders represent 26 and 30% of the total energy expenditure in fed and fasted hens. Differences in size and/or metabolic rate of these organs may have a considerably effect maintenance requirements. Interestingly, fasting increases liver and reproductive tract tissue metabolism in broiler breeders indicating the major role that liver plays in energy metabolism in fasten hens; and a compensatory response of magnum to an absence of dietary substrates for egg synthesis.

Reports on causes of variation in RFI and heat production in laying hens were examined by Luiting (1990). Major sources of variation were attributed to physical activity, feathering density, basal metabolic rate, area of nude skin, body temperature, and body composition. However, further studies using divergent selection have shown a great importance of body composition and lipid metabolism to explain RFI variation. While basal metabolic rate was found to be similar in high and low RFI lines, differences in feeding activity and regulatory thermogenesis were found. Variation in maintenance requirement may be attributed to differences in body composition. This can be affected by bird genetics, behaviour and management, and may affect fat and protein deposition in body tissues, lipid metabolism, egg composition and the size and metabolic rate of liver, gut and reproductive tract.

Heritability of RFI has been estimated in laying hens from 0.30 to 0.60. Schulman et al. (1994) looked for the genetic correlations of

RFI and economically important traits in laying hens, and only found a genetic correlation with feed consumption. Interestingly, when broiler stocks are provided a choice of protein and energy compared to a single, complete diet, they do not maximise their growth. Instead, they will grow more slowly with a reduced feed efficiency, and ultimately be more fat while also having an enhanced immune response. In the continuing push to grow broilers more efficiency, sight of what is 'normal' for the bird must not be lost in commercial stocks.

The newer broiler breeder genetic strains are becoming more specialised and appear to have more specific management methods associated with them. Effective ovary management is an integral part of a successful breeder management programme. Managing the broiler breeder female for optimal chick production requires an understanding of reproductive physiology, nutrition, and their interaction. With new analytical and descriptive tools to apply to daily broiler breeder management, the modern manager will be able to cope with the increasingly specific needs of the modern heavy breeder. By identifying 'reproductive attitudes' of individuals and their incidence in a population, more effective refinement of broiler breeder management strategies will be possible.

Broiler Breeders: Managing the Paradox between Reproduction and Growth

Modern poultry production is based on high broiler growth performance within a limited time frame. And at the same time obtain as many eggs from the parent stock as possible. These two do not seem to easily match, yet good reproduction results are within reach. But that demands adequate and precise management of the breeder flock.

In the world of broiler meat production, developments happen quickly. The increase in broiler performance in commercial breeds has been tremendous. This performance is not only established in growth per day, but also in feed conversion, carcass yield, breast meat yield, mortality, leg quality, etc. The modern broiler of today is barely comparable with the bird of 20 or 30 years ago due to very intense and very sophisticated genetic selection.

This is nicely demonstrated in a famous experiment in the USA. A group of researchers compared two lines of broilers, one with the genetic characteristics of the broiler lines of 40 years ago, and one with the characteristics of today. They also developed two different feeds, one with the characteristics of 40 years ago, and a modern broiler feed. When the different lines were given the different feeds, it became very clear that the improvement in performance is largely due to the genetic improvement, as the birds of today did well on the feed of 40 years ago, but the birds with the genetic profile of 40 years ago hardly benefited from the modern feed.

From an economical standpoint, it is very clear that a breeding company has no choice than to focus its selection programme mainly on broiler traits and to a much lesser extent on breeder traits. As the cost price of a day old chick is only a fraction of the cost price of a full grown broiler, a 10% improvement in broiler traits is worth much more money than a 10% improvement in breeder traits. After all, we are not in this business to produce hatching eggs or day old chicks, but to produce meat.

The Broiler-Breeder Paradox

Unfortunately, selecting on one trait has often consequences for other traits that are not necessarily directly associated with the trait that we select on. One clear example is the influence of selecting for growth on reproductive performance. It is often assumed that there is a negative correlation between broiler and breeder traits, sometimes called the broiler-breeder paradox. This paradox states that if the broiler characteristics in a line improve, its reproductive capacity will be impaired.

Figure 6: Despite very high genetic growth potential for broiler offspring, most commercial breeder flocks reach 85% peak production or even more.

Although from a biological standpoint this is true, our modern broiler lines paint a different picture. The genetic potential for growth and broiler characteristics has never been as high as now, however it is not uncommon for most commercial broiler breeder flocks to have

85% peak production or more. There are differences between the different breeds, but reaching 85 or even 90% peak of production is a level where 15 years ago producers could only have dreamed of.

We must have great respect for the geneticists that work in the different breeding companies, as well as for the research institutes and universities that have contributed to developing the tools that made these progresses possible, because they have been able to overcome the broiler-breeder paradox.

Modern Management

However, if we take a closer look at the different commercial lines, we see that the genetic improvements have an impact on the management strategies that we have to apply. As broilers are the end product of different lines with different characteristics, each breeding programme has to make a choice of which broiler traits are represented in which line. Although in reality all lines are selected on more or less all traits, the focus on which traits should be present in which lines is not equal for all breeds.

As a high growth potential in a female breeder has a negative effect on reproduction, some breeds focus their broiler traits more on the males than on the females. The result is that egg production in these lines goes very well, but the males with a higher growth potential need stricter management in rearing and production to have satisfying results. As they tend to get overweight more easily, it is more difficult to keep them active and in good condition then when the focus for broiler traits is more on the females.

If the females are genetically more selected on broiler traits, we see that female management is more difficult but a good fertility is easier in reach, resulting sometimes in flocks with 90% hatch of eggs set for a period of more than 10 weeks. Probably the biggest challenge for geneticists is to keep all the lines in balance with each other, in such a way that in the field the product can cope with all the circumstances that we can apply to them, as the same breed will be used at different places and in different conditions.

Modern poultry production is based on high broiler growth performance within a limited time frame. And at the same time obtain as many eggs from the parent stock as possible. These two do not seem to easily match, yet good reproduction results are within reach. But that demands adequate and precise management of the breeder flock.

Rearing Strategies

Although the broiler growth, and with that the body weights at for instance 42 days of age, have increased dramatically over the last decades, a quick scan of the breeder growth curves especially in rearing shows hardly any change over time. As the requirements for growth and maintenance have not really changed over the years, the amount of feed we have to give to realise the required growth curve has not changed dramatically either, but the amount of feed the bird could eat if it has the choice did increase tremendously.

Figure 7: The aim on the breeder side is to obtain as many good quality chicks per breeder hen as possible.

This means that the level of restriction that we have to apply on our rearing birds, as a percentage of the amount of feed they would consume when fed ad libitum, is gradually increasing. This increasing relative restriction requires more precise management, for instance to avoid competition and with it creating lack of uniformity.

Growing on Air

What we often see is that rearing flocks requires hardly any weekly feed increases in the period from 7-8 to 14-15 weeks of age to stay on body weight. It is as if in this period they grow from air. Of course birds do not grow on air, but because of their high eating capacity in the first weeks of life, we easily overfeed them in this period, and then it takes several weeks of hardly any feed increases to get them back on the correct body weight and growth curve.

This bears a risk in itself, as too severe feed restriction can result in negative effects such as delayed development at the end of rearing, which requires more feed and (over) stimulation at the beginning of lay.

Avoid Overfeeding

Due to the high growth potential of the birds, it is very easy to overfeed them. Especially just before and during the start of production, as well as towards peak production, overfeeding the flock will result in more formation of breast muscle, as that is where modern breeds are selected on. Although the extra breast muscle by itself doesn't necessarily have a negative effect on production, it results in a stimulation of sexual hormones, as the hormones that are associated with muscle growth are stimulating sexual hormones.

Figure 8: As broilers are the end product of different lines with different characteristics, each breeding program has to make a choice in which broiler traits are represented in which line.

As a result, some birds (especially the birds that are coming in production somewhat later) will start to produce more follicles than their system can handle, resulting in more double yolks, and if it is too severe in egg peritonitis, internal lay, poor peak and more mortality. Too avoid this, care must be taken not to stimulate the flocks too much with feed if the flock is not ready. After all, being selected so intensively for growth has resulted in a bird that exactly knows what to do with feed that it can't use for egg production...

Feed Reduction after Peak

Too much growth after peak leads to overweight birds and lack of persistency in production and fertility. With continuous selection on growth characteristics in the broilers, proper feed management during and after peak becomes increasingly important, to avoid excessive weight gains later on. Especially the feed reduction at and just after peak is an important tool to avoid overweight later in life, as at peak the birds need to grow much less than in the period coming towards peak.

But with the high peaks that our modern breeds obtain nowadays, it is difficult for a broiler breeder manager to decide to cut the feed when the birds are still at a very high level of production. This often results in a jump in body weight of 200-300 grams two to three weeks later, and with that higher body weight the birds have more risk of becoming overweight later in life.

Accept no Mistakes

Modern broiler breeds have a tremendous growth potential, and with it are still able to have a very good production on the breeder level. However, with the selection for broiler characteristics we do put pressure on our breeder lines, and this requires a good understanding of what the parent bird needs, and a strict management to fulfil these needs. With this high level of genetic capacity in our lines, we have to accept that no mistakes are permitted to achieve their maximum performance. Continuous adjustment and fine tuning of the management to deal with the continuous improvements in the birds is key.

3

Broiler Breeder Nutrition and Management

Genetics, Nutrition and Reproduction

Poultry breeding remains largely based on classical quantitative genetics. In essence, pedigree broiler candidates are full-fed nutritionally-dense and properly balanced diets to allow individuals that have the greatest potential to utilise crude protein (CP) and metabolisable energy (ME) to grow fast, convert feed efficiently, and yield well to become apparent by their performance. Thus, broiler strains are often selected on high-protein, high-energy diets.

Selection on nutrient dense diets apparently necessitates nutrient-dense diets in order for the progeny to fully express their genetic potential. An excellent example of the relationship between genetic progress and appropriate nutritional compensations can be taken from research with quail (Lilburn et al., 1992). Random-bred Japanese quail were placed on a selection programme intended to create heavy weight (HW) quail. These quail were full-fed 28% CP diets for 28 days and then the largest birds were selected and mated to produce the next generation. When these birds were reared to sexual maturity on a 24% CP diet, as recommended by the National Research Council (NRC, 1984), there was an obvious delay in sexual maturity (onset of egg production).

When the HW quail and the non-selected randombred control (RBC) quail were fed a range of diets differing in % CP from hatch to sexual maturity a nutritional-genetic interaction became evident. The RBC quail, when fed the NRC recommended 24% CP diet from hatch, matured sexually at about 42 days of age. In contrast, the HW quail exhibited delayed sexual maturity on the same diet. However, when the HW quail were fed a 30% CP diet, more like that fed during their pedigree selection process, the delay in sexual maturity was noticeably reduced. These data make the strong suggestion that declining reproductive function due to genetic selection for non-reproductive traits may in some way be ameliorated nutritionally.

Interaction of Nutrition, Temperature and Lighting Programme

The very important interaction between climate, photostimulation, and nutrition can be illustrated by examining the seasonality of broiler breeder reproduction in temperate climates. The differences in so-called "in-season" and "out-of-season" breeders have historically been attributed to daylength. However, the interaction between daylength and seasonal differences in temperature and feed intake provide an alternative explanation of seasonality. In-season breeders are generally the better performing birds in a temperate climate. These birds typically hatch in warm periods of the year when daylengths are long. Daylength and temperature both decline during the rearing period. As broiler breeders have typically been fed to achieve a body weight standard, the cool weather at the end of the rearing period dictates more feed be fed.

Thus, the cumulative nutrition is adequate for in-season breeders if photostimulation is not too early. In contrast, out-of-season birds hatch in the cool season and are reared while both daylength and temperatures are increasing. As the birds approach the age of photostimulation in warmer temperatures, they require less feed to achieve the standard body weight and thus have less cumulative nutrition at the point of photostimulation.

This causes a delay in onset of egg production and is frequently the case for tropical countries. Many managers respond to this with earlier photostimulation, but this often does not correct the problem. Increasing the target body weight has often been used as a "treatment" for out-of-season (hot temperature grown) birds because, as we now know, having a heavier body weight effectively increases the cumulative nutrition in the warmer weather.

Another method of correcting delayed onset of egg production in warm weather has been to delay photostimulation until sufficient cumulative nutrition has been achieved. With this latter approach, body weight will not become excessive, but this approach may not work as well as increasing the cumulative nutrient intake to 20 weeks of age. If the current genetic trend towards improved feed efficiency continues, breeders will have to be photostimulated much later and/ or grown to a higher body weight at 20 weeks of age in order to accumulate sufficient nutrition for proper responsiveness to photostimulation.

At this point, it should be stated that photostimulation plays a major role in the overall process of nutrient accumulation. Photostimulation somehow changes the birds from a "nutrient-accumulating" to a "nutrient-expending" organism. This is the probable reason that age at photostimulation has been delayed with good results in modern feed-efficient lines of broiler breeders.

An extended rearing period is needed for some birds to accumulate sufficient nutrition for optimum reproduction. As shown below, this is certainly true for females and one can interpret the large body of French literature to mean the same for males. In these male data, most heavy-line male fertility problems could be avoided by simply not photostimulating the birds and thus giving them unlimited time to accumulate sufficient nutrients necessary to sustain optimum reproduction before actually achieving sexual maturity. The act of photostimulation can obviously interrupt the process of nutrient accumulation.

The Concept of Minimum Cumulative Nutrition

During recent years, our laboratory has examined the relationship between cumulative nutrition during the rearing period and subsequent female reproductive performance. The rearing period was defined as the time from placement at one day of age to photostimulation at 20 weeks of age.

The groups were fed the same diet during rearing, but the feed was allocated differently each week to achieve the cumulative differences. There were apparently no great differences in female body weight, but when the birds were photostimulated at less than ~22,000 kcal cumulative ME and ~1200 g CP, there was a reduction in eggs per hen of ~15. This suggests that there was a minimum nutrient intake, irrespective of body weight, required to obtain acceptable levels of egg production.

A recent review of NCSU broiler breeder research flock data revealed that in 1988, females were grown to a 140-day body weight of ~2.0 kg with ~28,000 kcal cumulative ME. Comparative data from

1998 shows that this 2.0 kg body weight could be achieved with as little as 20,000 kcal cumulative ME. This difference is probably due to the remarkable genetic progress made in broiler feed conversion. This may explain why photostimulation has been required to be adjusted from 126 days in 1983 to 154 days or later today. With improved feed conversion, it may simply take longer to accumulate the necessary nutrition for a proper response to photostimulation.

Fertility in the Female

The fact that cumulative CP nutrition at photostimulation can have a significant effect on female fertility has been clearly defined. The female contributes to fertility through mating receptivity and spermatozoal storage in special sperm host glands in the oviduct. This was demonstrated by VanKrey and Siegel (1974) where broiler line genetic selection proceeded on nutrient-dense broiler diets while typical lower protein and energy rearing diets were used for parent stock. Evidently, inadequate CP (amino acid) nutrition prior to photostimulation, irrespective of female body weight, leads to poor persistency of fertility.

The latter is a good indicator of persistency of female fertility as all males were managed in a similar manner across all experimental groups. It is also important to note that the effects of nutrition and management during rearing and the early breeding period are often seen only very late in the breeding period. This projected minimum assumed that the total lysine, on a corn-soy-based diet, was 5% of crude protein and methionine + cystine were 83% of lysine.

Feeding Programmes for Yield-Type Broiler Breeders

It has been noticed in the USA that females reared with males often produce more eggs than females reared sex-separate. In order to understand this observation, a study (Mixgrow) was conducted to determine the effect of mixing males with females at different ages. Yield males were fullfed on an 18% CP diet until mixed with females at two, four, six, or eight weeks of age. The yield females received an 18% CP diet for one week followed by a 15% CP diet to photostimulation. The female feeding programme used was one that had been shown to be successful for the "standard" type of broiler breeder pullet. Female body weights were virtually identical across male treatments.

These fertility numbers are lower than optimum because males and females were fed together after 21 weeks of age to exaggerate the effect of cumulative nutrition during rearing and to allow the males to be exposed to a decreasing feed allocation after 35 weeks of age. In spite of this, some of the pens with the eight-week mixed males exhibited fertility in excess of 90% at 64 weeks of age without any body weight control or separate feeding.

The later mixing age males (six and eight weeks) were more resistant to the feed reduction after peak egg production because they reached sexual maturity with a greater nutrient reserve. The actual feed intake of the males mixed with females at six weeks of age (as an example) and that of the females can be estimated from the body weights taken from all birds every two weeks using the formulas of Combs (1968).

The males consumed about 125% to 150% of the female feed intake depending upon age when mixed and body weight. This would give an actual cumulative ME intake of over 34,000 kcal and 1600 grams of CP for both the six week and eight week mixed males. This agrees with other data from our laboratory with separate-grown males. The data also show that the real pattern of female feed consumption differed significantly from the programmed pattern, especially after 14 weeks of age.

This must be extremely important as females that were grown sex-separate on the programmed female feed amounts laid ~35 fewer eggs per hen. These data (and field experience) suggest that larger feed increases late in rearing (in blackout where there is little reproductive development) for "yield-type" pullets results in excessive body weight and excessive "fleshing" (breast meat development). Much has been said about the need for good "fleshing" in "standard" strains of parent stock but the situation is much different for the "yield-type" pullet. Excess breast meat appears to reduce egg production.

We must be careful to not give too much feed too early (before onset of lay) as we may simply increase female body weight, primarily breast meat, and cause reproductive problems such as peritonitis. The excess breast meat probably increases maintenance and inhibits reproductive development. This may be why heavy breasts relative to fat pad develop when feed increases are too rapid in "yieldtype" females. These birds with excess breast meat relative to fat pad tend to exhibit a reduced appetite in hot weather (even in tunnel-ventilated and evaporatively cooled houses), increased susceptibility to heat stress, poor peak egg production and lay poorly thereafter. A conservative feeding approach both before and after photostimulation would be advisable with "yield" females until one becomes familiar with the particular strain of broiler breeder in the specific situation. It is better for the hens to be late coming into production than to exhibit high mortality and poor egg production. These problems are uncommon with a "standard" type broiler breeder hen.

In a manner similar to the need to modulate any large increases in feed intake, diets should be formulated to minimise abrupt changes in composition that will create situations that are similar to abrupt changes in the feeding rate. A single dietary ME for all diets is recommended to assist production managers maintain consistent feed increases. Similarly, modern broiler breeders may respond robustly to abrupt changes in protein with an unexpected increase in body weight. A smooth transition among starter-grower-breeder diets or starter-grower-prebreeder-breeder diets should be considered during feed formulation.

It is suggested that total lysine levels be ~5% of crude protein and methionine + cystine be ~0.60-0.63% of the diet for most feeds. It is probable that "yield-type" females perform better with a slightly lower protein breeder feed than can be fed successfully to a "standard" female. A 16% CP diet with ~0.80-0.82% total lysine should be sufficient to support egg production without producing excessive amounts of breast meat.

Dietary Protein and Metabolisable Energy for Broiler Breeder Males

Few data exist that link intake of ME during rearing to breeding performance. However, the findings of Vaughters et al. (1987) indicated that a relationship between ME consumed during rearing and fertility may exist. Our data suggest a minimum cumulative ME intake of ~30,000 kcal prior to photostimulation. However, most data suggest that reproductive fecundity is directly related to daily ME intake during the breeding period and that daily ME intake should somehow be proportional to body weight and body weight gain. It should be stated that Parker and Arscott (1964) and Sexton et al., (1989b) observed that decreased fertility was preceded by decreased dietary ME intake during the breeding period. In cages, Attia et al. (1995)

fed Ross males 300, 340, or 380 kcal ME per day. They found no fertility differences, but did note increasing testis weights with increasing ME intake. In floor pens from 26 to 60 weeks of age, Attia et al., (1993) found the 300 kcal ME males to weigh less and have lower fertility than the males consuming 340 and 380 kcal ME per day. These data clearly show a differential effect of ME intake in cages versus floor pens due to the difference in relative activity levels. All the birds in cages probably received enough ME to satisfy their reproduction requirements. However, in the floor pens, it appeared that the birds on the lowest ME intake did not receive enough nutrients for reproduction due to the increased maintenance requirement required for increased activity.

It is also very interesting that these authors found a dose-related decrease in 42-day broiler weights with decreasing ME allocation to the breeder males. Presumably, these data suggest that males that have the potential to produce the largest broilers require more ME to breed in natural mating conditions. These data also suggests that excessive efforts to control male body weight can reduce broiler performance.

Confusion about optimum diets for males began when Wilson et al. (1987a) fed 12%, 14%, 16%, and 18% CP diets to males from four to 53 weeks of age. The 10 males used per treatment were placed in cages at 14 weeks of age. There was no lighting programme detailed in the manuscript and is presumed to be natural daylight during rearing with artificial supplementation at some unspecified point. Cumulative CP to 21 weeks was 1220 grams and 1385 grams, respectively for the 12% and 14% groups. This total increased to 1650 grams at 27 weeks of age for the 12% group, the time of the first artificial ejaculations in this particular study.

By comparison, males in natural mating conditions need to mature by ~22 weeks of age for best results. No significant differences in semen volume, testis weights, and spermatozoal concentration among the diets were found, but significantly more males produced semen as a result of abdominal massage on the 12% and 14% CP diets. Although there were no significant differences in body weight among the treatments, the 12% and 14% males did exhibit a generally more consistent body weight gain throughout the breeding period. It is important to note that all the diets used in this and subsequent studies from this laboratory at Auburn University had total lysine as

5.1% to 5.3% of total CP and total methionine + cystine as 75% to 77% of lysine in corn-soy based diets. This was similar to the dietary approach used by our laboratory at North Carolina State University, but may differ somewhat from observed commercial practice where low protein male diets are often not properly balanced. We like to have lysine as 5% of CP and methionine + cystine in the range of 75% to 83% of lysine.

In a recent study from the same laboratory at Auburn University, Zhang et al. (1999) made a comparison of 12% and 16% CP diets from four to 52 weeks of age. As in previous reports, there was a higher percentage males producing semen as a result of artificial ejaculation, but there were again no differences in semen quality or quantity. Given that differences in semen quality or quantity are not usually found as a result of difference in CP intake, one has to question if the reported higher percentage males producing semen as a result of artificial ejaculation is simply an artifact of the semen collection process with birds that may vary in body conformation.

This response (percentage males producing semen) seems to consistently take the form of a dose response while all other variables show no such dose response. In the experiment of Zhang et al. (1999), the daily ME allowance was 325 kcal during the breeding period. As shown later, this energy allocation is slightly low. A gradual decline in semen production with increasing age and body weight was observed, irrespective of CP level of the diet. The authors interpreted this to mean that continued body weight gain was necessary to maintain optimal male reproductive function. Continued body weight gain clearly would require appropriate increases in ME allowances as body weight increased.

The extensive French work, led by de Reviers showed that heavy weight line males exhibit greater problems with persistency of testes size and semen production when compared to medium weight male lines. Photostimulation of heavy weight line males typically result in a robust, but short, response in testicular weight and semen production while medium weight male lines exhibit better persistency of these traits. It is presumed, as no nutritional data were given in these reports, that both male lines were fed typical low-density diets. The problem of lack of persistency of semen production can be solved, if one is using artificial insemination, by simply not photostimulating the birds and allowing the males to reach sexual maturity at their own pace, presumably after consuming sufficient nutrients.

Therefore, if a bird were deficient in CP during the growing period the effects would be most noticeable around the onset of sexual maturity. Vaughters et al., (1987) fed diets containing 12%, 15%, or 18% from 24 to 27 weeks of age (early breeding period) and reported initial fertility to be highest for the 18% CP diet in natural mating conditions. This suggests a relationship between sexual development and the initiation of reproductive function. Turkey and broiler breeder hens are both known to exhibit an intense desire to mate prior to the onset of egg production.

When turkey hens were inseminated during this period of prelay receptivity, there was a significant increase in life-of-flock fertility even in the presence of marginal spermatozoal numbers (McIntyre et al., 1982). This early mating presumably leads to enhanced spermatozoal storage. This may also be true for broiler breeders. It is clear that broiler breeders that exhibit low initial fertility under commercial natural mating conditions, where sexual maturity is needed at about 22 to 24 weeks of age, have difficulty achieving optimum fertility at later ages.

Although there appears to be an impact of CP during the growing period on fertility during the breeding period, dietary CP appears to have less impact during the breeding period. Diets from 5% to 16.9% CP have produced similar results in cages. The reason that these previous workers did not see more differences in fertility due to breeder dietary CP was probably due to the fact that their experiments were often initiated later in the breeder period (after 28 weeks of age). In these experiments, it appears that the birds were not marginal in CP before the experimental diets were applied, which made it difficult to detect fertility differences due to differences in breeder dietary CP. These data also suggest that low protein male feeds should not be used before sexual maturity is complete.

Data from our laboratory suggest the minimum cumulative CP intake required prior to photostimulation for broiler breeder males involved in natural mating to be on the order of 1600 grams, as compared to the 1200 grams required for female. We have found that it is possible to achieve this nutrient target with diets ranging from 12% CP to 17% CP. Moreover, our data, shown below, demonstrate the interaction of body weight and feeding programme that influence male reproduction so profoundly. The broiler breeders were the Ross 308 package but the data are illustrative of our data with Cobb 500 and Arbor Acres Yield broiler breeder packages as well. All of these birds were reared separately from the females and fed sexseparate during the breeding period.

The effect was more pronounced for the 17% CP males that were slightly larger and evidently less resistant to the imposed feeding deficiency. The problem was corrected by a five grams increase in daily feed allocation for the males. The cumulative intake of nutrients at 21 weeks of age were 1568 g CP and 36,593 kcal ME for the 12% males and 2123 g CP and 36,593 house was at 23 weeks of age. Again, the data suggest that if the minimum nutrition is adequate, it is not important what dietary protein level is used to achieve the goal. The males with the most consistent body weight gains produced the best fertility.

Body Weight in Broiler Breeder Males

It has long been clear that feed restriction to control body weight is both obligatory and beneficial in broiler breeders. However, excessive feed restriction of males during part or all of the growing period has been associated with decreased early fertility (Lilburn et al., 1990). Based upon the discussion above, it is thought that this effect is due to insufficient cumulative nutrition at photostimulation.

The major impetus for sex-separate feeding during the breeding period was the observation that poor fertility was associated with overweight males and separate feeding was believed to be necessary to control male body weight. However, caged males fed near ad libitum are known to exhibit excellent spermatozoal production. This suggests that an appropriately controlled feed allocation rather than severe restriction is required.

It is likely that overly severe feed restriction has actually caused fertility problems due to reduced mating activity as a result of caloric deficiencies. This may help explain the observations of Hocking (1990) who performed experiments with males in floor pens with natural mating during the breeding period. He found a curvilinear relationship between body weight and fertility. This implied that if body weight were too low or too high there would not be optimum fertility. He observed that underweight males were not physiologically sufficient while overweight males often were physically incapable of completing the mating process. He suggested an optimum body weight for optimum fertility that changed with age. He concluded that restricted control of body weight should allow an increase in body weight with age of the male.

We conducted a study to examine this inconsistency. We found that a decrease in fertility coincided with a decrease in female feed allocation and an increase in male body weight in situations where males were fed with females. In a similar manner, a decrease in male feed allocation in situations where males and females were fed separately caused a transient decrease followed by an increase in male body weight coincident with a decrease in fertility. Thereafter, fertility again increased when the feed allocation was increased in the separate-fed males. Male body weight was better controlled and fertility improved when the male feed allocation was increased slowly rather than decreased.

What can be the explanation for the paradox shown above? As an example, a typical male at ~30 weeks of age will weigh ~4.00 kg (8.8 lbs.). The daily maintenance requirement at ~21°C (70°F) is ~306 kcal while that of a 4.45 kg (9.8 lbs.) male at ~45 weeks of age would be ~329 kcal. Unless there has been an increase in daily feed allocation proportionate with the body weight gain, the 4.45 kg male would have to exhibit negative growth (lose body weight) as the male mobilised body reserves to make up the energy deficiency.

This would continue until the energy reserves of the larger male were exhausted. At this time, mating activity would decrease as testosterone levels decreased. The male would then gain body weight because of inactivity. This could lead one to conclude that males do not necessarily cease mating because they gain excessive body weight, but that males gain excessive body weight because they cease mating!

In the same way that our best egg production occurs when the females slowly gain body weight, our best fertility occurs when the males slowly gain body weight. As the male does not exhibit a decline in daily energy requirement as does the female (due to decreasing egg production) it is suggested that the daily feed allocation be increased at least one gram every three to four weeks during the breeding period such that the male body weight increases slowly but consistently and remains within limits established by practical experience and known to be associated with good fertility. Ken Krueger (1977) found that male turkey semen production could be maximised for the entire life cycle by maintaining the toms on a feeding regimen that supported a consistent weekly body weight gain. Any loss in body weight was associated with a decline in semen production.

Broiler Breeder Male Mortality

The mortality of "yield-type" broiler breeder males during the laying period has become a costly problem for the USA poultry industry. The average male mortality from 22 to 64 weeks during the years 1995 to 1999 was approximately 43% (AgriStats, Inc., 6510 Mutual Drive, Fort Wayne, IN 46825). The cause of the majority of this mortality is unknown. To test a theory about the cause of this high mortality, Ross 308 females and non-dubbed Ross males were raised sexseparate on either a "linear" or a "concave" feed allocation programme.

One group of males was reared with females on a "mixed" programme. Birds were grown on a daily 8-hour light and 16 hour dark lighting programme and both feed and water were controlled. At the end of 21 weeks, the birds were moved to a curtain-sided laying house and photostimulated. "Linear" grown males received constant feed increases of 2.4 g per male/week from four to 28 weeks. After 28 weeks, males received a constant feed amount of 117 g (342 kcal ME) per bird. All separate grown males received the same amount of cumulative feed through 21 weeks that resulted in a cumulative CP intake of 1600 g and a cumulative ME intake of 32,000 kcal per male at photostimulation at 21 weeks of age.

Separately grown males on the "linear" programme had significantly higher body weights than separately grown males on the "concave" programme. However, there were no differences in body weight due to treatment after photostimulation (22 weeks). All males had similar body weights at 22, 26, 28, 40, and 52 weeks of age.

It appeared that the "mixed" grown males had less mortality during this period although these differences were not statistically significant. Males grown separately on the "linear" programme or "mixed" with females had significantly higher mortality from 30 to 44 weeks when compared to males grown separately on the "concave" programme. From 45 to 64 weeks, "linear" males numerically had the highest mortality with "mixed" and "concave" males having similar mortality. When mortality was compared from 30 to 64 weeks, "concave" males had significantly lower mortality when compared with the "linear" males. "Mixed" males were intermediate. This same trend was observed overall (22-64 weeks). All data indicated that males grown on a "linear" feed allocation programme exhibited higher mortality than males grown on a "concave" feed allocation programme. It appears that the majority of the mortality due to "linear" feeding can be expected to occur between 30 to 44 weeks of age.

It appeared that 117 g of feed (342 kcal ME) per male per day was adequate to keep males grown on a 17% CP diet slowly gaining weight from 28 to 60 weeks of age under our current research management that utilises strict male and female exclusion grills. Fertility was excellent with these males.

However, under commercial conditions, a gradual increase in male feed allocation would be advised. Therefore, it appears that the feed allocation programme used during the growing period in association with the time of photostimulation can influence broiler breeder male mortality. It appears that this occurs irrespective of body weight. The various groups of males employed in these experiments exhibited average body weights that were not remarkably different. Thus, one can conclude that management of feeding programmes should take limited precedence over body weight management.

Overview of Separate Male Rearing

Males may successfully be reared separately from females throughout the growing period. Careful attention to the feeding programme must be exercised as demonstrated by the field observations outlined below. There has been much discussion about optimum male BW at four and 20 weeks of age. A thorough examination of all available data suggest a minimum required cumulative nutrient intake from day old to photostimulation of ~1600 g CP and ~32,000 kcal ME per male and a specific feeding programme approach is required to minimise mortality and maximise fertility.

We can conclude that the amount of nutrients a bird has available throughout its life impacts fertility and egg production. Metabolisable energy (or feed allocation) available to the bird during the breeding period is directly correlated to fertility, egg production and body weight. Protein accumulated in the bird during rearing influences the age of sexual maturity and the level of initial fertility for both males and females. Dietary CP has the largest impact during the grower and pre-breeder periods, as this is when most of the CP required for initial sexual development is accumulated.

Body weight and house temperature need to be controlled within certain limits throughout the life of the flock, however temperature and body weight management is most critical late in the breeding period because body weight is greatest at this time. The data clearly show that no specific diet has more or less utility for a male broiler breeder. Diets ranging from 12% CP to 17% CP can be fed provided that the cumulative intake of CP to photostimulation is sufficient to support initial sexual development.

However, it is clear from practical experience that changing from a moderate or high CP feed to a low CP feed prior to sexual maturity has adverse effects on broiler breeder fertility. It is most important to maintain consistent body weight gain throughout the life of the broiler breeder. Abrupt increases or decreases in body weight are clearly associated with changes in fertility and egg production.

This infers a need to closely align ME intake to maintenance requirements that are driven by body weight and temperature. In summary, all the rules for the "standard" broiler breeder remain basically true, but more attention must be paid to these details. The most obvious exception to the basic rules is that excessive "fleshing" can be detrimental in the yield-type female because it can increase sensitivity to environmental temperature, reduce egg production and appetite, and increase mortality.

Special Notes and Acknowledgements

Some estimates of metabolic energy requirements in the text were based upon the formulas of G. F. Combs, 1968, the Proceedings of the Maryland Nutrition Conference for Feed Manufacturers. These estimates have been found to be reasonably accurate, but may need to be adjusted slightly for strain and age effects and should be used with some caution.

Portions of this manuscript were excerpted from the Proceedings of the Poultry Beyond 2005 Conference held in Rotorua, New Zealand in February 2001 and from the Proceedings of the Australian Poultry Science Symposium held in Sydney in February 2001.

Feed Processing –Impacts on Nutritive Value and Hygienic Status in Broiler Feeds

Most of the technological processes in compound feed manufacturing impact on nutritive value and hygienic status of the resulting feed and sometimes act synergistically, (e.g. expansion plus pelleting on salmonella decontamination). The nutritive effect is mainly exerted on feed digestibility and efficiency of energy utilisation. The return on energy invested in feed processing (electrical and steam energy), is positive for the majority of processes when balancing against the gain in available feed energy.

In terms of dietary protein (amino acids) and feed additives like vitamins, enzymes and probiotics, special attention must be paid when applying thermal processes. Only a feed/purpose specific selection of treatment processes and fine-tuned process parameter setting brings optimal results in a scenario with sometimes conflicting objectives: maximum improvement of nutritive value combined with optimal decontamination of pathogenic germs. Whereas an improved nutritional status of the processed feed is durable, pre-cautionary measures must be taken to maintain its hygienic status.

Domestication and breeding of animals has improved animal performance in an unprecedented manner. The requirements in terms of nutrients have increased accordingly. Feedstuffs as offered by nature cannot meet these requirements, even if they are offered in a variety, without further processing. The coverage of dietary energy needs poses a particular challenge in feed formulation and subsequent manufacture. In the process of making dietary nutrients available for metabolic purposes the digestion process is a major influencing factor.

Digestive capacities of animals have remained largely unaffected despite centuries of breeding and selection (Wenk, 1982). Therefore, the preparation of feedstuffs before ingestion to improve digestibility became a major focus in nutrition research. On the other hand, the production units for livestock have become larger driven by economy of scale. This has heightened the importance of hygienic regimes including feed hygiene to prevent disease outbreaks and securing hygiene and feed ingredients of animal origin.

It can be stated that short and long conditioning have very little effect on nutritive value. Grinding and crumbling have no hygienic effect. Conditioning with subsequent pelleting has some and expansion plus pelleting has pronounced effects on both nutritive value and hygienic status. Toasting is reserved for the treatment of single ingredients like soybeans, peas, beans, canola and others. It is used to eliminate or reduce the content of anti-nutritional factors in these ingredients, before they can be used in compound feed. From that angle this process is very significant in terms of improving the nutritional value of these ingredients, which otherwise could be used only with limitations. The toasting process is not primarily applied for microbial decontamination, however, the processing conditions would allow for this.

Grinding

Poultry have a short digestive tract and therefore the digestibility must be high; but is capable of grinding entire grain in the gizzard. Nir and Hillel (1994) found a correlation between particle size and weight of gizzard and duodenum in 21days-old broilers. Also the gizzard pH-value is significantly lower when the feed particles are coarser. Feeding partly entire grains thus is seen as a possibility to prevent intestinal infections in poultry. Kamphues et al. (2005) made similar observations in weaning piglets when testing the effect of potassium diformate on Salmonella in faeces. The number of animals excreting Salmonella and the duration of excretion was significantly lower, when the feed was coarse ground (control diet: 32% > 1.0 mm; 26% < 0.4 mm; test diet: 58% > 1.0 mm; 10% < 0.4 mm). The authors concluded that Salmonella shedding is related to feed structure.

In broiler feeding, pelleted feed is the preferred choice whereas in layers its mash feed. Finely ground mash increases the time for feed intake and reduces feather picking (Walser, 1997). Particle size structure should be uniform to prevent nutrient imbalances by selective feed intake. An optimal mash structure can be achieved with a combination of expansion and crumbling process. It also must be mentioned that the level of fineness of mash impacts on subsequent processes like pelleting (throughput, pellet quality). Grinding of feed ingredients is a pre-requisite of mixing different ingredients and achieving a low coefficient and variation percentage of nutrients in the feed mash.

Conditioning and Pelleting

All processes require the use of a short-term conditioner for steam and water addition. Pelleting agglomerates smaller feed particles with the help of mechanical pressure, moisture and heat to larger particles. This tends to improve animal performance due to less feed wastage, no selective feeding, and improved palatability and starch gelatinisation. Numerous trials have shown better daily gain and feed conversion broiler feeds.

Average daily gain and feed conversion ratio is improved by 5-8% and 3-5% respectively, when feeding a pelleted diet with zero fines versus a mash diet. Pelleting has an effect on the hygienic status of the feed; however, a short-term conditioner alone, or in connection with a pelleting press, will not be sufficient for decontamination of pathogenic microorganisms. Already Friedrich (1979), Hacking et al. (1978) and Pietzsch (1985) have stated that pelleting is not sufficient for safe decontamination as relatively large numbers of microorganisms remain in the finished feed. Sufficient decontamination with pelleting as the final processing step is achieved, when using a hygieniser with horizontal retention screw or vertical shaft (similar to long term conditioner or ripener), assuring minimum retention times of three minutes. It is suggested to increase retention time at the technically highest possible moisture level; however, limits apply for the proper functioning of the pellet press (< 16% moisture).

Expansion

Expansion of mash feed is an established thermal processing technology that is widely used for broiler diets to improve pellet quality, include higher liquid and fat levels and increase operational productivity by increasing pelleting line capacity, and to enhance the flexibility of ingredients usage and animal performance (Wilson et al., 1998). This technology simultaneously leads to achieve a hygienic status, described below as "commercial sterility".

Most studies document an improvement in body weight gain and feed conversion ratio; however, depending on the selection of ingredients, enzyme supplementation to expanded diets proves to be obligatory to achieve this. Gauer (2002) reported no differences in FCR, when energy level in a corn-soy broiler diet (ME 3200 kcal) was decreased by 3.1%.

Separate treatment of single ingredients can prove beneficial for some but not all ingredients.

The results show that expansion is not acting on all ingredients in the same manner. In corn-soy diets an increase in dietary energy improves feed conversion ratio. In wheat or barley based diets, the addition of enzymes is recommended to reduce the gut viscosity that is inevitably increased by expansion of these ingredients. Only then a positive effect on body weight gain and feed efficiency is achieved.

Energy Return

The increase in metabolisable energy for pelleting and expanding and pelleting has been measured in several broiler trials. In pelleting, energy expenditure (57 kWh/t) and gain in ME (56 kWh/t) are about the same. Thus for pelleting the technical energy expenditure is recovered by the increase in metabolisable energy of 1 – 1.5%. In expanding and expanding plus pelleting a net gain in feed energy versus technical energy is observed. About 73 kWh/t are expended in return for 122 kWh/t extra feed energy. As trials have shown, the ME increase in corn-soy-based broiler diets can exceed 4%.

These calculations consider only the nutritional improvement by enhancing the digestibility at equal feed intakes. Better economic feed conversion ratios by reduced feed losses (dust, fines, uneaten feed) will further improve the value of these technological processes. A total energy balance should include energy for transportation and storage processes of feed. High-density feeds resulting from pelleting after expansion are advantageous in terms of transport and storage space and feed intake in broilers.

Hygienic Feed Preparation

Decontamination

Sterility cannot be achieved in feedstuffs; however, a so-called "commercial sterility" (Asquith, 2002) is possible. This means that pathogenic microorganisms have been eliminated (coliform bacteria, Salmonella, moulds, etc.). A higher, but still incomplete level of sterility can be reached at temperatures of 130°C, pressures of 3 bar and treatment times of 20 minutes, as practiced when sterilising meat-meal (autoclaving). This technology is reserved for special applications because valuable components, such as amino acids and vitamins, are damaged at these temperature and pressure conditions.

Appropriate processing technology depends on proper judgement of how these organisms live, grow and die.

Salmonella, for example, have the highest growth rate at a temperature of 35-38°C. But, they can also grow at temperatures between 5-50°C if the ambient conditions are optimal. This depends mainly on the moisture content. Salmonella can only reproduce at an aw-value (available water) of >0.92, which is not found in normal feedstuffs with ~13% moisture as it requires a moisture level of >25%. The aw -value is the part of the water in feedstuffs, which is not bound to other substances but is completely available to microorganisms. This value varies between 0 (anhydrous substance) and 1 (pure water). It indicates the equilibrium moisture, adjusted between the sample and the relative air humidity.

Salmonella and coliform bacteria need an aw value of >0.92 for growth, while moulds need >0.8. The aw value of cereals with 17% moisture is ~0.8. To avoid spoilage, the cereals needs drying to <16% moisture. The temperature range allowing growth of Salmonella, other bacteria and moulds is in the range of 5-55°C. When heating the product to more than 60°C, microorganisms stop growing and die. They do not die abruptly, but according to a logarithmic function.

Apart from aw values and temperatures, the mortality rate also depends on the pH value.

Salmonella and other bacteria have good growth conditions in the pH range of 7.0 to 8.5, moulds from 5.0 to 7.0. The reduction of pH value by adding organic acids can be used to decontaminate feeds. For complete decontamination, the addition of about 2-4% of organic acids is required. The costs involved are much higher when compared with thermal treatment; also, special attention is needed for the selection of acid-resistant equipment. In dry feed at ambient temperatures, Salmonella cannot reproduce, but they do survive by downscaling their energy metabolism. When the aw -value increases, the energy metabolism rekindles. Activated, moist salmonella can be eliminated much easier than non-activated Salmonella. For this reason, hydrothermal treatment should always be coupled with an increase in moisture.

Salmonella are not uniformly distributed in the feed, but are found in spots throughout the mixture. Therefore, it is difficult to "locate" them and to account for them in samples. For this reason coliform bacteria counts are used as a measure for the decontamination effect of different treatments since they exhibit the same heat resistance as Salmonella and are ubiquitously and uniformly distributed in feedstuffs. Some organisms, such as spores of aerobic bacteria and anaerobic sporoform bacteria (Clostridium perfringens), cannot be eliminated readily by thermal processes. The same applies for toxins formed by moulds. Heidenreich (2002) showed that moisture content of about 14% in expanding at 105 °C is needed for a significant reduction in total germ counts. Israelsen et al. (1996) and König (1994) aimed at similar conclusions. In expander processing prior to pelleting the expander peak temperature must clearly exceed 100°C to result in total elimination of Salmonella. In expanders and extruders, next to temperature and moisture, the sudden pressure drop at the outlet is a key element in killing bacterial cells, causing the living cells to burst (Peisker, 1991). Without pelleting, the expander temperature must reach 110°C for total Salmonella elimination.

Prevention of Recontamination

Recontamination is a true hazard in feed production. A multitude of steps must be taken to secure the hygienic status achieved by feed processing. The addition of organic acids, such as propionic or formic acid (~ 0.5%) is generally accepted to protect a clean feed mixture after hygienic treatment against recontamination. As a consequence, the design of the cooling area for pellets, the finished product sector, the bulk transport to the farm and the storage of the feed in the farm silos must be taken into account for maintaining hygienic status of finished compound feeds.

Moulds were completely eliminated and total germ count considerably reduced after thermal treatment. The weak spot at the feed plant level are the ducts from the pellet press to the cooler and the cooler itself. Such equipment in particular should be addressed in HACCP-programmes. Also unloading pits, dust filters, elevator legs, bins and trucks must be checked on regular basis. In the farm silo the feed maybe completely re-contaminated due to feed residues and insufficient cleaning before recharge.

Fully integrated poultry companies can address this and establish a successful Salmonella control system, comprising raw materials, feed processing (broiler, parent stock and layer feed), transport systems and farms. For example, the German poultry integrator "Wiesenhof" has adopted a policy enabling them to offer guaranteed Salmonella free hatched chicken, broiler meat, table and breeding eggs (Gill, 2002). However, in the commercial feed industry sector similar policies can be agreed upon contractually and several "tracking and tracing" systems have been developed by private or governmental entities and are quickly gaining importance in the industry.

Role of Broiler Breeder Genetics on Breeder Chick Quality and Sensitivity to Overfeeding

Broiler breeders must have the genetic potential for efficient growth as well as the ability to effectively reproduce. However, the interaction between nutritional and reproductive traits is complex and continually changing with the introduction of new genetic lines. These studies were designed to gain a better understanding of how selection for growth traits and variability among strains affects growth traits, reproductive morphology and production traits of broiler breeders. Breeder strain influenced sensitivity to photostimulation, to excess feeding and to pullet growth profile. By understanding how sensitivity to changes in nutritional status differs among strains differing in muscling, then feeding and management strategies can be refined to maximise the production efficiency of the hen.

Broiler breeders are a moving target. While broiler 42-day body weight is increasing each year, the target body weight for mature male and female broiler breeders has changed little. In 1979, Hubbard male and female breeders were approximately 50% of the 42-day broiler weight. In 2001, this percentage had decreased to 36.1 for males and 30.3 for females.

The situation for knowing what nutrients these birds need is compounded by the development of "yield" lines, carrying increased amounts of breast muscle yield, often on a smaller carcass frame. This increased growth efficiency is expressed partly in the greater capacity for muscle growth. Pym et al. (2004) indicated that differential fractional rates of protein deposition, breakdown and synthesis have resulted in increased protein retention in a high compared to a low efficiency line.

The reproductive efficiency of broiler parents is increasingly dependant on very specific feed restriction and lighting programmes to optimise reproduction. Broiler breeders must have the genetic potential for efficient growth as well as the ability to effectively reproduce. Excess body weight can result in reduced egg production, hatchability, liveability, egg weight, feed efficiency and increased shell porosity. Overfeeding can accelerate the sexual maturation process and elevate ovarian large yellow follicle (LYF) numbers in birds of similar BW. Furthermore, feeding programmes during rearing and early lay can change frame size and breast muscle fleshing in the birds.

These studies were designed to gain a better understanding of how selection for growth traits and variability among strains affects growth, conformation, composition, reproductive morphology and production traits of broiler breeders.

Strain Variation in the Acceleration of Sexual Maturation

Broiler breeder strains vary in the extent to which sexual maturation is influenced by nutrient intake (Robinson et al., 1998). Four commercial strains were reared on a common body weight target, and fed one of three feeding programmes from photostimulation (22 wk): ad libitum; Fast-Feed (weekly feed adjustments based on a 5 g increase for every 5% increase in production); and Slow-Feed (daily adjustments of 1 g/d between 22 and 26 weeks of age, and 0.5 g/d until 31 weeks of age).

There was a one-week range in the mean age at first egg among the strains. Ad-libitum feeding did not accelerate sexual maturation in two of the strains, suggesting these birds have a later maturation of the hypothalamo-pituitary axis. This is a significant finding that serves to show a basis for genetic differences in photo-sexual response among commercial stocks. In these late-maturing strains, it would seem to be pointless to subject these birds to increasing day lengths and feed allocations as soon as the operator would more traditional early ¬maturing strains. In the strains responding to overfeeding, maturation was accelerated by 6 to 7 d, on average.

The number of large yolky follicles varied among strains, with the two strains that reached sexual maturity first having the fewest large follicles. These data strongly suggest that it is essential to follow management recommendations specific to a breeder genotype.

Impact of Growth Selection on Sensitivity to Overfeeding

A study was designed to show the effects of degree of selection for yield traits on the ability to cope with a feeding challenge. The strains were: Random-bred (unselected since 1977), Ross 308 (a high-yield bird suited for the whole-bird market), and Ross 508 (a very high-yield bird suited for the cut-up and further processing market). Each strain was raised to the same target body weight at 20 wk of age, when they were individually caged. Beginning at photostimulation (22 wk of age), pullets were fed 100% (control), 120%, and 140% of the feed needed to maintain the Ross 508 growth curve. A total of 90 birds were dissected at sexual maturity and 144 were kept to 58 wk of age for measurement of production traits.

The timing of sexual maturity was affected by strain, with the RB20, Ross 508 and Ross 308 birds laying eggs 16.5, 20.2, and 27.4 d after photostimulation. Birds of all strains appear to have acquired the appropriate level of growth and composition to support rapid sexual maturation. At sexual maturity (onset of lay), the Ross 508 birds had the highest proportion of breast muscle. Conversely, the Random-bred hens were the fattest – reflecting the less efficient growth of their older genetics. The 120 and 140% treatments only added an additional 5.3% and 9.7% to BW at sexual maturity, respectively.

Feeding regimen had a big impact on egg production, with 166, 159, and 137 settable eggs produced by the 100, 120, and 140% groups, respectively. Interestingly, the modern, high breast-yield Ross 508 birds were the most sensitive to overfeeding, producing 177 eggs with the 100% feed allocation compared to only 123 eggs on the 140% feed allocation. When coupled with decreased rates of fertility and hatchability in the 140% treatment, overfeeding had a devastating effect on chick numbers.

The hatchability of the 140% Ross 508 hens near the end of the study ranged between 20 and 30%. At the end of the trial, 94% of the 100% feed allocation birds were still in active lay compared to 86% in the 120% allocation group and only 63% in the 140% allocation group. The productivity of the Ross 308 hens was least impacted by feed allocation, demonstrating a better tolerance to a range of feeding profiles than either the Random-bred or the Ross 508 hens in this study.

Effect of Strain and Growth Profile on Production Traits

An experiment was performed to test how the interactions between genetic strain, age at photostimulation and target body weight profile impact growth rate and efficiency, nutrient partitioning, sexual maturation and reproductive efficiency.

The strains used were: Hubbard Hi-Y, Ross 508, and Ross 708. The four body weight profiles separated at 5 wk and converged at 32 wk of age as follows: STANDARD (control); LOW (12 wk body weight target = 25% lower than STANDARD followed by rapid gain to 32 wk); MODERATE (12 wk body weight target = 150% of STANDARD followed by lower rate of gain to 32 wk); and HIGH (12 wk BW target = 200% of STANDARD followed by minimal growth to 32 wk).

One of the primary effects of the growth profiles was on frame size. The long-term concern would be that feeding a small-framed to the same body weight target, as a larger framed bird will result in increased fatness and the triggering of reproductive disorders associated with overfed hens. During the period immediately after photostimulation at 18 wk, the LOW birds had a very high feed allocation relative to that of the other growth curve treatments to allow their weight profile to converge with the others by 32 wk.

Despite what would normally be considered excess feed, sexual maturation was still delayed in the LOW birds. In contrast, the HIGH birds had a very low feed allocation during this period, which delayed sexual maturation the Ross 708 hens and suggesting that these birds do not tolerate nutrient shortages well. These birds carried a greater proportion of breast muscle and less fat than the other strains, which may contribute to their inability to cope with reduced feed at this critical time. Photostimulating birds at 22 wk of age alleviated most of these problems.

Body weight at sexual maturity was 3.40, 3.21, 3.01, and 2.84 kg for HIGH, MODERATE, STANDARD, and LOW birds, respectively. The body weight differences impacted shank and keel length, indicating differences in frame size. Interestingly, ovary weight in the later maturing LOW birds (55 g) was 6 g heavier than in the other groups. The number of large yellow follicles on the ovary did not change with photostimulation age (average of 7.5 follicles), except in the HIGH birds, where it dropped from 8.1 in birds photostimulated at 18 wk to 6.5 in those photostimulated at 22 wk. Feed allocation to the HIGH birds was quite low during this period to keep the body weight on target, which likely impacted ovary development.

The reduced feed on the MODERATE and HIGH profile also reduced early egg size and stunted the length of the prime sequence (the characteristically long daily egg laying sequence occurring early in lay). Although these birds were larger, their early production traits were similar to that of a much smaller bird. This illustrates how recent feeding level may have a greater impact on production traits than body or growth pattern does.

Ultimately, the 18 wk PS-age birds yielded 8 more eggs (170) than 22 wk PS-age birds to 58 wk of age, with no affect on unsettable egg production. On average, total egg production was similar among growth profile treatments. However, there was variability in the productivity of specific strains grown on some profiles. The Ross 708-HIGH hens, for example, under-performed (138 eggs) compared to the other profiles (mean = 166.3). Alternatively, Ross 508-HIGH birds laid the same number of eggs as Ross 508-STANDARD birds (mean = 178.7).

Examination of individual growth profiles revealed strain-based strategies for managing reproduction. The Ross 708 tied up nutrients deposited during the pullet phase tightly, and was unable to mobilise nutrients from storage, as they were needed under conditions of dietary deficiency. This may be partly due to their increased breast muscle mass. Under more normal feeding conditions these birds performed very well.

The Hubbard Hy-Y hens appeared much more able to mobilise nutrient stores, and were not hindered by the very low feed allocations provided to the HIGH birds during sexual maturation. Economic analysis of production traits revealed that using a STANDARD feeding profile and photostimulating pullets at 22 wk of age was most often the best management practice on a cost/chick basis. Photostimulating birds at 18 wk of age resulted in higher total egg production, although much of this advantage was lost in small, unsettable eggs (<52 g) and a higher degree of production variation among hens. Ultimately feeding profiles affected egg production traits differently among strains, with little effect of photostimulation age.

The newer broiler breeder genetic strains are becoming more specialised and appear to have more specific management methods associated with them. Both genetic strain and feeding treatment affected how the birds came into production and had some influence on carcass fleshing traits. However, this did not have a consistent effect on egg production traits. The negative effects of overfeeding were more pronounced in the highest breast-yield strain. These studies indicate some of the complexity in the interaction between nutritional and reproductive parameters and demonstrate the need for strain-specific management strategies.

Daylength for Broiler Breeders – Have We Got it Right?

From the start of the broiler industry in the 1950s, recommendations for lighting breeders have invariably been based on programmes designed for lighting egg-laying hybrids. Typically, these involve an initial period of long days followed by a step-down programme to reach 8 hours by 1-2 weeks of age. The birds are held on 8 hours until about 20 weeks and then transferred to a 10-, 11- or 12-hour day to initiate egg production, followed by a series of weekly increments to reach a maximum of 15 or 16 hours.

The reason for the dependence on egg-type lighting programmes has been the minimal amount of research into the broiler breeder's response to light, presumably because it was assumed that it would respond similarly to its lower bodyweight, egg-type cousin. However, in recent years fundamental research conducted at the University of KwaZulu-Natal in South Africa has shown that broiler breeders cannot simply be regarded as large chickens and may, in some aspects, be better treated as if they were small turkeys. So what have we got right with broiler breeder lighting, and where do we need to make changes?

Controlled Environment in the Rearing Period

The most important finding of the recent research has been that broiler breeders, unlike egg-type hybrids, still retain a vestige of seasonal breeding. Indeed, surveys of performance in two broiler breeder industries (Mexico and Spain) have shown that despite flocks being grown to typical breeder company bodyweight targets and lit according to their recommendations, they still show a pronounced seasonal variation in the age at which they come into lay.

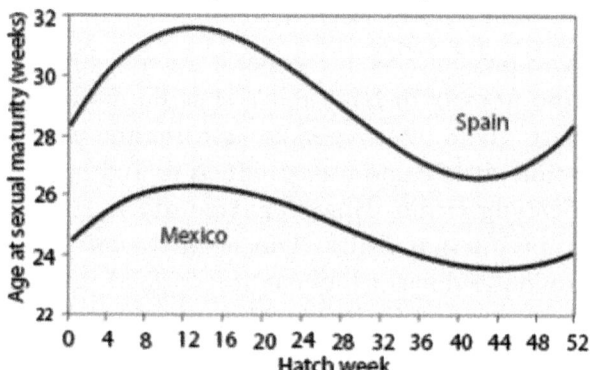

Figure 1: Effect of hatch date on sexual maturity in broiler breeders kept in the northern hemisphere (subtract or add 26 to produce hatch dates for southern hemisphere flocks)

It will be seen that the range is wider in Spain because of the bigger difference between the shortest and longest days at higher latitudes. A characteristic of seasonal breeding birds is that they are unable to respond to a stimulatory daylength when they hatch. Instead, they require a period of short days before they can respond to a transfer to longer days. As a consequence, spring-hatched birds, which have never received less than 12 hours light, are generally the latest flocks to mature. Those that hatch in the autumn will be the earliest to start egg production because they will have experienced the short days of winter during the rearing phase. In nature, this inhibitory mechanism, termed photorefractoriness, stops birds coming into lay in the autumn. This is because food will be scarce, and cold winter conditions will reduce the progeny's chances of survival.

Turkeys are seasonal breeders and typically need about two months of short days (mimicking winter) at the end of the rearing period to make them responsive to a light increase. However, in contrast to turkeys, broiler breeders are control-fed during rearing. This not only slows up their growth but also delays the dissipation of the refractory condition. As a result, they require more like 5 months of short days before they can be successfully stimulated into egg production with a light increase.

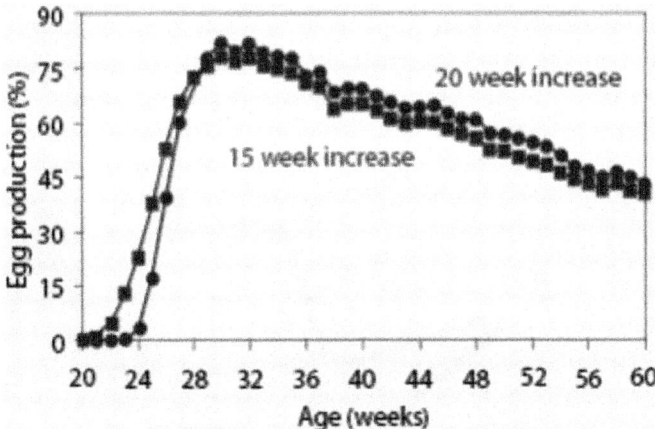

Figure 2: Egg production for 2.1kg body weight broiler breeders given an increase in daylength at 15 or 20 weeks

Our research in South Africa has also shown that photostimulating broiler breeders much before 19 weeks – even if feed restriction during rearing is relaxed – is likely to result in a flock of birds being sexually uneven and having a low peak, poor persistency, and variable egg weight. The egg production data shows the slow entry into egg laying and low peak for birds photostimulated at 15 rather than 20 weeks of age, even though each group was grown to achieve the same bodyweight (2.1kg) by the time of the light change.

There is little difference in the age at which egg production begins when birds are reared on 6, 8 or 10-hour days and transferred to 16-hour days at 20 weeks. So it seems that there is little wrong with the current recommendation to rear broiler breeders on 8-hour days prior to a transfer to long days at about 20 weeks of age.

Table 1: Performance Of Broiler Breeders On 6, 8, or 10 Hour Daylengths and Transferred to Sixteen Hours at Twenty Weeks

Daylengths During Rearing (Hours)	Age At 50% Egg Production (Days)	No Of Eggs To 60 Weeks Of Age	Average Egg Weight (g)
6	196	161	68.1
8	193	169	67.5
10	196	171	68.4

Non-Lightproof or Open-Sided Housing in the Rearing Period

Contrastingly, the advice to rear spring-hatched birds that are kept in open-sided or non-lightproof housing on a daylength equal to the longest anticipated natural daylength is incorrect, even though this continues to be an appropriate lighting policy for rearing commercial egg layers.

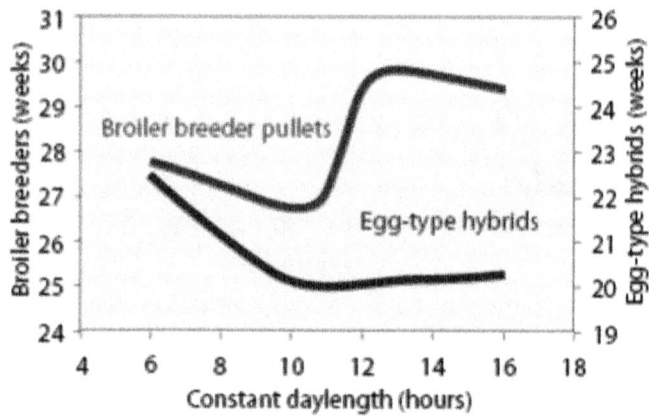

Figure 3: Age at 50% egg production in broiler breeder and egg-type hybrids kept on constant daylenghts

Broiler breeders that are never exposed to short days take much longer to become responsive to an increase in daylength. Indeed, those reared on long days take about 3 weeks longer to start egg production than birds reared on 8-hour days and never photostimulated. This is completely opposite to the response of egg-laying birds. A further problem when broiler breeders are given long days during the rearing period is that it leaves little opportunity to photostimulate them when they are moved into the laying house.

In one of the trials in South Africa, we compared three different lighting programmes during the rearing period:

• a naturally increasing daylength from either 10 or 11 hours to 14 hours

• a decreasing schedule from 14 hours down to 10 hours, or

• a constant 14-hour daylength.

There was only 3 days difference between the various treatments in the time taken to reach 50% egg production after they had all been transferred to a 16-hour day at 20 weeks. This demonstrates the dominant role that the feeding programme plays in the control of sexual maturity in broiler breeders.

However, the failure to provide the 14-hour birds with any short days meant that this group had very poor persistency; some birds will have gone out of lay, and overall, they laid 6-10 fewer eggs to 60 weeks than the groups reared on the simulated natural lighting programmes. For seasonal breeding species that are held on long days, the greater the delay in the start of egg production means they go out of lay sooner at the end of the breeding season.

Figure 4: Performance of Broiler Breeeders Reared on Simulated Naturally Increasing or Decreasing Daylengths of Constant 14 Hour Days before a Transfer of 16 Hour Days at 20 Weeks

Lighting Programme To 20 Weeks	Age At 50% Egg Production (Days)	No. Of Eggs To 60 Weeks Of Age	Average Egg Weight (g)
Increasing From 10-14 Hours	209	150	170.0
Increasing From 11 To 14 Hours	210	146	69.8
Decreasing From 14 To 10 Hours	209	150	69.4
Constant 14 Hours	212	141	68.9

Controlled Environment in the Laying Period

Egg production over the laying cycle when broiler breeders are given a single increase from 8 to 16 hours at 20 weeks of age is inferior to that when they are transferred to only 12-hour days. Data trial conducted in South Africa show that birds transferred to 16 hours came into lay slightly earlier than 12-hour birds, but had poorer egg production after peak, which resulted in the production of 10 fewer eggs to 60 weeks of age.

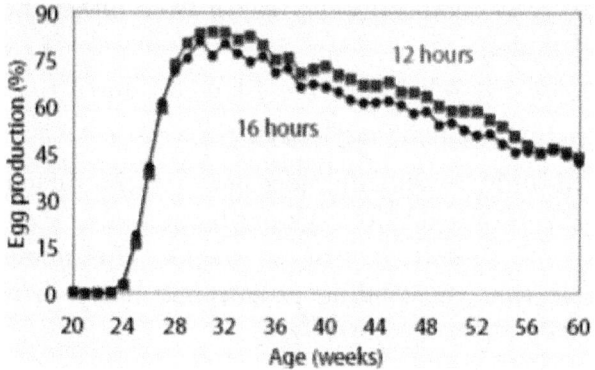

Figure 5: Egg production for broiler breeders abruptly transferred from 8 to 12 hours or from 8 to 16 hours at 20 weeks

Other birds were transferred to 12 hours at 19 weeks then given four further increases to reach 16 hours at 23 weeks. Their egg numbers to 60 weeks were similar to those for birds changed to 16 hours in a single increase at 19 weeks.

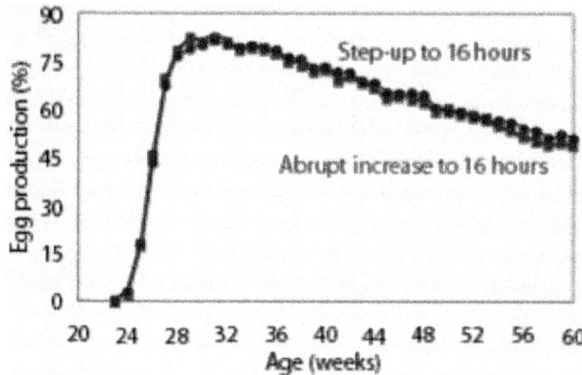

Figure 6: Egg production for broiler breeders given a single increase from 8 to 16 hours at 19 weeks or an initial increase to 12 hours then a series of further increases to reach 16 hours at 23 weeks

However, there were differences in the pattern of egg production, as occurs in egg-type hybrids: birds that were abruptly increased came into lay slightly earlier and had a higher peak rate of lay than birds that had been given the gradual increase. However, the latter group had better egg production at the end of the laying period.

These findings suggest that broiler breeders do not need to be given 15 or 16-hour daylengths (as recommended by most breeding companies); a single transfer to 12 hours will be adequate for maximum egg production.

Laying hens require 1% more energy per hour when they are in light than when they are in darkness. When broiler breeders are given a fixed daily amount of feed and 12 hours of light, they will have 4% more energy available for egg production than birds given 16 hours light.

Non-Lightproof or Open-sided in the Laying Period

It is probably sensible to give broiler breeders a 16-hour day in the laying period when they are kept in non-lightproof accommodation. This gives more opportunity to photostimulate (albeit minimally) birds when they are transferred to the laying house in mid summer, and avoids a decrease in daylength after the longest day.

Lighting of Males

Research has yet to be concluded on the response of males to lighting. Preliminary results suggest that males respond to a lighting programme in a similar way to females. That is, they also require a 4-5 month period of short days during rearing and should not be photostimulated much before 20 weeks of age if uniform sexual maturation is to be achieved.

What's wrong...what's right?

- The recommendation to rear broiler breeders that are kept in controlled environment housing on 8-hour days and transfer them to 11 or 12 hours is sound.
- The breeders' advice to photostimulate at 20-21 weeks is absolutely correct. This ensures that, within a flock that has been grown to reach 2.0-2.2kg by 20 weeks, all birds will be able to respond to an increase in daylength.

- When birds are reared in non-lightproof (open-sided) accommodation, it is probably better to expose them to naturally changing daylengths than to rear them on a constant long day.
- It is unnecessary to continue increasing the daylength to 15 or 16 hours in a lightproof laying house; 12 hours is perfectly adequate for maximising egg production.
- 16-hour daylengths need to be given to birds kept in open-sided or non-lightproof housing at high latitudes, but this can be reduced to 14 or 15 hours to match the longest natural daylength at lower latitudes.

4

Limiting Ovarian Development to Maximise Chick Production in Broiler Breeders

There are several important instances in the management of broiler breeders where "more" is not "better". In simple terms, "more feed" allotted to hens to the extreme degree of ad libitum feeding reduces settable egg production by as much as 40 eggs per hen.

Breeder hens use dietary energy differently than egg-type hens. Young broiler breeder hen respond to extra dietary energy by developing extra ovarian follicles. Several research projects have been carried out at the University of Alberta, that have shown that "more ovarian development" does not mean "better reproductive efficiency". Clearly, a high priority during the early laying period is to promote ovarian development at a rate that will result in the highest number of settable eggs. This paper will look at the process of sexual maturation of the breeder pullet to improve egg production, fertility and hatchability.

The critical time in broiler breeder management is the period from photostimulation (lighting) to peak production. This period is characterised by relatively fast weight gains, and the changes brought about by the hormones being produced by a newly active ovary. Photostimulation is generally considered the cue to initiate puberty, although the response to light is readily modified by level of feeding.

Light energy passes through the skull and illuminates the hypothalamus. When the hypothalamus receives a photostimulatory signal (long day length above a threshold of intensity), hormones involved in ovarian function are produced. One of the first responses you can see by looking at the ovary of the bird after lighting is that the very small ovarian follicles begin to increase in size. The small follicles produce large quantities of estrogen, the hormone that causes most of the reproductive transformation associated with puberty.

Firstly, estrogen influences the production of yolk precursors (building blocks of yolk production)in the liver of the bird. The liver visibly enlarges and becomes paler as its fat content increases for the production of egg yolk lipids.

Secondly, the oviduct increases in size so it is ready to receive ovulated follicles. This step appears to be a limiting one for heavy turkeys, as the ovary in these birds develops faster than the oviduct does, so there is a high incidence of follicles that are ovulated that are not picked up (internal ovulation) and hence, not formed into normal eggs.

Thirdly, estrogen influences changes to bone composition that allow calcium to be mobilised daily to make egg shells.

Finally, estrogen combines with male sex hormones (androgens) which results in changes to plumage, comb size and colour and sexual receptivity to males.

Shaver Starbro pullets were reared to 20 weeks of age to determine the influences of lighting programme and feed allocation from 20-25 weeks on ovarian development and production traits.

The experiment used two lighting programmes and two levels of feeding. A "slow photoperiod" treatment (SP) was gradually brought from 8L:16D with weekly increases in day length. A "fast photoperiod" treatment (FP) was increased from 8L:16D to 15L:9D in a single step at 20 weeks of age. The "slow feed" treatment group (SF) received very moderate increases in feed allocation from 20 to 25 wk of age. Increases in feed allocation of greater than 5 g, were divided up into two smaller increases per week.

The "fast feed" treatment group (FF) received a more generous feed increase between 20-25 weeks of age. It was not the original intent to decrease feed allocation to these birds; however, at 22 wk of age we discovered that the increases at 20 and 21 weeks of age had a greater effect on body weight that planned. The decision to decrease the feed allocation at 22 weeks of age was made to prevent the fast feed hens from becoming too heavy. It should be noted that the withdrawal of feed did not result in a decrease in body weight in subsequent weeks. From 25 to 64 weeks of age all of the birds were exposed to the same photoperiod (15L:9D) and a common feed allocation.

A group of birds were killed on the day following sexual maturity to examine development of the reproductive tract and carcass composition. Individual egg production records were kept and eggs were incubated at weekly intervals to assess treatment effects on chick production. Feed allocation was decreased after peak egg production to keep the birds at proper body weight targets (for age).

Body Weight: There were no significant differences in age at sexual maturity between any of the groups. Body weight and fat pad weights at sexual maturity were not different in terms of main effects. Within the group of slow photoperiod hens, the fast feed programme resulted in heavier body weights compared to the slow feed programme.

Ovary: Ovary morphology was highly influenced by both the photoperiod and feed allocation treatments. The slow photoperiod treatment had a greater ovary weight than did the fast photoperiod treatment. The fast feed treatment resulted in significantly heavier ovary weight and one more large follicle than the slow feed treatment. To our knowledge this is the first report of a relatively minor feeding difference affecting large follicle recruitment in commercial breeders.

Egg Production: All treatment groups had very high rates of total egg production when expressed in terms of main effects or the interactions of the main effects. There was a 10.9 egg advantage to use of the slow feed feeding programme compared to the fast feed programme. This fact is presumably related to improved control of the follicular recruitment process, as fast feed birds also had fewer large follicles at sexual maturity.

Fertility and Hatchability: Hatchability was reduced in the fast photoperiod programme compared to the slow photoperiod programme. Embryonic mortality was examined on day 7, 14, and 21 of incubation. The greatest embryonic mortality was found in the fast photoperiod treatment in the later stages of incubation.

Eggs from the fast fed hens had higher embryonic mortality than eggs from the slow fed hens. This indicates that developmental problems may be associated with the eggs from birds with excessive follicle development. These data suggest that minor differences in feed allocation have an influence on ovarian form and function. Proper management of feed allocation during the pre-lay phase is critical to Optimise the number of saleable chicks produced.

The above research data support the hypothesis that chick production in broiler breeder hens can be compromised by over-feeding early in lay. Future management programmes for breeder hens should focus on limiting excessive follicle production so that there is an orderly cooperation between the ovary and the oviduct in the production of hatching eggs.

Such a programme would minimise egg shell problems, reduce the incidence of multiple-yolked eggs and result in long egg-laying sequences. Long egg-laying sequences lead to high peak production and excellent persistency of lay. Further studies on limiting follicular growth by controlling energy intake and where nutrients are partitioned or distributed through the body and the role of light in sexual maturation are underway at the University of Alberta.

Broiler

A broiler is a type of chicken raised specifically for meat production. Modern commercial broilers, typically known as Cornish crosses or Cornish-Rocks are specially bred for large scale, efficient meat production and grow much faster than egg or traditional dual purpose breeds. They are noted for having very fast growth rates, a high feed conversion ratio, and low levels of activity. Broilers often reach a harvest weight of 4-5 pounds dressed in only five weeks.

They have white feathers and yellowish skin. This cross is also favourable for meat production because it lacks the typical "hair" which many breeds have that necessitates singeing after plucking. Both male and female broilers are slaughtered for their meat. In 2003, approximately 42 billion broilers were produced, 80% of which were produced by four companies: Aviagen, Cobb-Vantress, Hubbard Farms, and Hybro.

History

Before the development of modern commercial meat breeds (cows, chickens, etc.) broilers consisted mostly of young male chickens (cockerels) which were culled from farm flocks. The males were slaughtered for meat and the females (pullets) were kept for egg production. Compared to today, this made chicken meat scarce and expensive compared to eggs, and chicken was a luxury meat. The development of special broiler breeds decoupled the supply of broilers from the demand for eggs. This, along with advances in nutrition and incubation that allowed broilers to be raised year-round, allowed chicken to become a low-cost meat.

Broilers are often called "Rock-Cornish," referring to the adoption of a hybrid variety of chicken produced from a cross of male of a naturally double breasted Cornish strain and a female of a tall, large boned strain of white Plymouth Rocks. This first attempt at a hybrid meat breed was introduced in the 1930s and became dominant in the 1960s. The original cross was plagued by problems of low fertility, slow growth, and disease susceptibility, and modern broilers have gradually become very different from the Cornish x Rock hybrid.

Modern Variants

Access to a special diet of high protein feed delivered via an automated feeding system. This is combined with artificial lighting conditions to stimulate growth and thus the desired body weight is achieved in 4 - 8 weeks, depending on the approximate body weight required by the processing plant. After processing, the poultry is delivered as fresh or frozen chicken to the stores and supermarkets.

Figure 1: Five day old broiler strain Cornish-Rock chicks

Because of their efficient meat conversion, broiler chickens are also popular in small family farms in rural communities, where a family will raise a small flock of broilers.

Broilers are sometimes reared on a grass range using a method called pastured poultry, as developed by Joel Salatin and promoted by the American Pastured Poultry Producers Association.

The term "broiler" is widely known in North America, Australia and England but not elsewhere in the English speaking world. The term "broiler chicken" is very widely used in Pakistan and India, as it was in the former German Democratic Republic and still nowadays in some eastern parts of Germany. The term is also used in Bangladesh, Indonesia, Sweden, Nigeria, Finland, Poland, Turkey and the Balkans.

Broiler Health Issues

Broiler chickens may develop several health issues as a result of selective breeding. Broiler chickens are bred to be very large to produce the most meat per animal. The large chickens cannot stand because their bodies grow too quickly for their legs. Therefore, they may become lame or suffer from broken legs. Broiler chickens are also prone to heart attacks for the same reason, as the heart cannot support blood flow to the large body of the chicken. Another issue with selective breeding is the larger chickens have a more aggressive appetite. The broilers are feed restricted and this leads to behavioural issues in chronically hungry birds.

Broiler chickens may often get joint disorders because their legs cannot bear the heavy bodies. A Swedish study by SLU Skara (Swedish farming university) revealed that only 1/3 of studied broiler chickens that were about to be slaughtered were healthy. Additionally, it is very inactive and as a result is a poor forager, prone to predation, and is generally not suited to small free range homestead flocks.

If the litter in the pen is not properly managed to prevent birds from standing and resting in their feces, painful hock burns and foot ulcerations and blisters can occur. Pastured birds which are rotated frequently typically do not have these issues.

Leptin Receptor in the Chicken Ovary: Potential Involvement in Ovarian Dysfunction of Ad Libitum-fed Broiler Breeder Hens

The ovary of the mature hen contains a hierarchy of yellow yolky follicles and several thousand smaller follicles from which the large yolky follicles are recruited. The yellow follicles are arranged in a size hierarchy and are committed to ovulation. In each follicle, the granulosa cells are surrounded by theca tissue and are separated from it by a basement membrane. Each compartment of the largest yellow follicles (theca and granulosa cells) can be anatomically separated to follow the individual functions of these two ovarian compartments in follicle growth and differentiation. At an early stage of follicular development the small ovarian follicles produce estrogens and androgens. As follicles begin to sequester yolk their production of estrogens from theca cells decreases to become very low at ovulation. As follicles are recruited into the yolky follicular hierarchy, estrogen and androgen production by theca cells diminishes and production of progesterone by granulosa cells increases. The largest F1 yellow follicle then attains the highest progesterone production at the time of ovulation.

In most domestic animals, reproductive function is considerably affected by nutrition. Various hormones, including growth hormone, insulin-like growth factors (IGFs) and insulin, have been proposed as potential mediators affecting reproductive function. However, the interactions between the reproductive endocrine axis and the metabolic axis have not been clearly determined. Leptin represents also a good candidate for such reproductive-metabolic interactions. Leptin, the protein hormone synthesised and secreted mainly by adipose tissue, has primarily been shown to regulate food intake and energy expenditure. Recent studies have demonstrated that leptin may also be involved in the regulation of reproductive mechanisms in human and rat ovaries.

Exogenous leptin can rescue reproductive function in ob/ob leptin-deficient mice that are not only obese but also infertile. This leptin action is independent from weight loss since feed restriction in ob/ob female mice fails to restore fertility. Leptin can also advance the onset of puberty or at least reverse the delay caused by feed restriction in rodents. In chickens, leptin attenuates the negative effects of fasting on ovarian function.

Injections of leptin during fasting delays cessation of egg laying, attenuates regression of yellow hierarchical follicles, alters ovarian steroidogenesis and limits apoptosis. Leptin exerts its effect by binding to a receptor which belongs to the cytokine receptor super-family. The chicken leptin receptor has been cloned and sequenced. Its expression at the level of the ovary suggests that leptin might act directly on the ovary to regulate chicken reproductive function.

Standard broiler breeders have been submitted to high selection pressure on growth and feed efficiency. Male traits have been favoured resulting in poorer reproductive performances of the hens. As a result of such selection, broiler breeder hens are subject to metabolic disorders and reproductive dysfunction. Overfeeding during reproductive development is associated with the formation of excessive numbers of ovarian yellow follicles which can be arranged in multiple hierarchies, with increased production of unsettable eggs. Severe feed restriction during rearing reduces the production of yellow follicles, the incidence of double ovulation and considerably improves the laying rate. Up-regulation of yellow follicles has been related to excessive recruitment and rapid growth rate of follicles to maturity, especially under ad libitum feeding. However, the mechanisms that regulate these processes are still not fully explained.

This study investigated the potential involvement of leptin and its receptor in ovarian abnormalities observed in broiler breeder hens fed ad libitum. We aimed to determine the evolution of leptin receptor gene expression in both granulosa and theca cells from the four largest yellow follicles of 32 week-old hens during the laying period. The effects of genotype and diet on plasma leptin levels and ovarian expression of the leptin receptor gene were also quantified.

For this purpose standard broiler breeder hens fed ad libitum or feed-restricted were compared to a French "Label" genotype and a dwarf "Experimental" line. Compared to fast growing standard broiler breeders, the French "Label" is a dwarf, slow-growing broiler genotype and the "Experimental" line is a dwarf genotype with a growth potential of progeny chicks close to that of the standard broiler chicks. The "Experimental" line is specifically selected for reproductive traits and viability at partial expenses of growth performances. The "Label" line tolerates ad libitum feeding of breeders and does not have reproductive problems under ad libitum feeding whereas the «Experimental» line can be fed ad libitum on a low energy diet during the growing period but presents more reproductive problems than the Label genotype but less than the standard genotype fed ad libitum.

Methods

Animals: Three lines of broiler breeder hens supplied by Hubbard primary breeder (Chateaubourg, France) were used in this experiment. The S line is a standard fast growing broiler, the French Label (L) line is a slow growing broiler breeder strain used for the quality market and the experimental (E) line is a broiler breeder strain bearing the "dw" dwarf gene. This strain has a decreased need for rationing. S and L hens were given the same regime in accordance with Hubbard nutritional recommendations (2724 kcal/kg) in fine meal form. During the growing period (0–20 weeks), half of the S hens were feed restricted (SR) on the same diet in order to match a reference body weight curve provided par Hubbard primary breeder, the other half (SA) and all the L hens were fed ad libitum.

The feed intake was equivalent to 37% of the SA group up to point of lay. A special diet was designed for the E hens, consisting of a series of finely ground meal diets with a lower energy content (2550 kcal/kg). The interest of the E group is mainly practical, it represents an actual alternative for severely restricted standard broiler breeder hens. Transition between the grower and breeder feed occurred at 20 weeks of age for the ad libitum-fed hens (L, SA and E). Transition occurred at the beginning of laying for the restricted hens that were then allowed ad libitum access to food.

At 32 weeks of age, blood samples were collected from 12 hens of each experimental group (SA, SR, E and L) and six hens were sacrificed by an overdose of pentobarbital (Sanofi-Sante Animale, Libourne, France). The ovaries and liver were immediately removed. Granula and theca compartments from the first (F1), second (F2), third (F3) and fourth (F4) largest ovarian yellow follicles were dissected as previously described.

Since SA birds presented a greater average number of yellow follicles per ovary (9.36, 8,00, 7.42, and 6.33 yellow follicles/ovary for the SA, SR, E and L hens respectively) and a higher proportion of pairs of yellow ovarian follicles undergoing simultaneous development, follicles were assigned to the same follicular rank when the difference of weight between two follicles was of 0.4 g or less. In that case, one follicle per pair was collected and dissected. Tissues were immediately snap frozen in liquid nitrogen and stored at -80°C until used for total RNA extraction. This experiment was carried out with due regard to the legislation governing ethical treatment of animals, and investigators were certificated by the French government to carry out animal experiments.

RNA Extraction and Leptin Receptor RT-PCR

Total RNA was extracted from liver, granulosa and theca cells using RNA InstaPure (Eurogentec, Angers, France) according to the manufacturer's recommendations. After DNAse treatment using Ambion's DNA-free kit (Clinisciences, Montrouge, France), 2 µg of total RNA were reverse-transcribed (RT) in a final volume of 20 μl using RNAse H- MMLV reverse transcriptase (Superscript II, Invitrogen, Cergy Pontoise, France) and random hexamer primers (Promega, Charbonnières, France). cDNA was then diluted to 1:8.

For normal PCR amplification, five microliters of the RT reaction were amplified for 35 cycles in a 50 μl reaction volume containing 2.5 units of Taq DNA Polymerase (Amersham Biosciences, Orsay, France), 2.5 mM $MgCl_2$, 0.2 mM dNTPs (Promega, Charbonnieres, France) and 0.2 μM of each forward and reverse primer. Leptin receptor forward (5'-GTC CAC GAG ATT CAT CCC AG-3') and reverse (5'-CCT GAG ATG CAG AGA TGC TC-3') primers were chosen according to the previously determined sequence of the chicken leptin receptor cDNA.

This pair of primers amplifies a 271 bp cDNA fragment located in the coding sequence of the extra-cellular domain. The amplification conditions were as follows: denaturation at 94°C for 30 sec, annealing at 60°C for 30 sec and primer extension at 72°C for 60 sec. After final extension at 72°C for 10 min, PCR products were resolved on 1.5% agarose gel containing ethidium bromide.

Real-Time RT-PCR

Real-time RT-PCR was performed as previously described. Briefly, forward leptin receptor primer 5'-GCATCTCTGCATCTCAGGAAAGA-3' and reverse leptin receptor primer 5'-GCAGGCTACAAA CTAACAAATCCA-3'(nucleotides 362 to 448 of the chicken leptin receptor cDNA sequence) were designed to be intron-spanning to avoid co-amplification of genomic DNA using Primer Express Software (Applied Biosystems, Courtaboeuf, France). A 20 μl master mix containing 12.5 μl SYBR Green PCR Master Mix, 1 μl forward primer (300 nM), 1 μl reverse primer (300 nM) and 5.5 μl water was prepared to perform real-time PCR (Applied Biosystems, Courtaboeuf, France). Five microliters of cDNA dilution was added to the PCR Master Mix to a final volume of 25 μl.

The following PCR protocol was used on the ABI Prism 7000 apparatus (Applied Biosystems, Courtaboeuf, France): initial denaturation (10 min at 95°C), followed by a two-step amplification programme (15 sec at 95°C, followed by 1 min at 60°C) repeated 40 times. Quantification was performed using ABI integrated software as previously described. 18S ribosomal RNA was chosen as the reference gene. The level of 18S RNA was determined using the Pre-developed

TaqMan Ribosomal RNA control kit (Applied Biosystems, Courtaboeuf, France) according to the manufacturer's recommendations. The results were expressed as the leptin receptor mRNA/18S RNA ratio. Each PCR run included a no template control and replicates of control and unknown samples. Runs were performed in triplicate.

Plasma Lipid, Glucose and Hormone Analyses

Total cholesterol, phospholipid, and triglyceride plasma concentrations were calculated with "Cholesterol RTU", "Phospholipides Enzymatique PAP 150", and "Triglycerides Enzymatique PAP 150" kits (bioMérieux, Charbonnieres les Bains, France) according to the manufacturer's recommendations. Plasma glucose levels were measured by the glucose oxidase method using an automated analyser. Plasma insulin levels were determined by a radioimmunoassay with a guinea pig anti-porcine insulin antibody using chicken insulin as the standard. Plasma concentrations of leptin were determined by a multi-species leptin RIA kit (LINCO Research Inc, CliniSciences, Montrouge, France) according to the recommendations of the manufacturer.

Statistical Analysis

The results of plasma lipid, glucose and hormone analyses as well as leptin receptor mRNA expression in the liver were analysed by one-way ANOVA and means were compared by Student Newman Keuls multiple comparison test. The effects of the groups of birds (E, L SA and SR), follicular rank and possible interaction on the logarithm of leptin receptor mRNA levels were tested by two-way ANOVA using the General Linear Model (GLM) procedure of SAS (SAS Institute, 1999. SAS User' Guide, Version 8 ed. SAS Institute Inc., Cary, NC). An additional effect of the subject was introduced into the model in order to take into account the fact that measurements of leptin receptor mRNA expression for the different follicular ranks were performed on the same animal. Pairwise comparisons of means for each significant effect of the ANOVA were performed by Scheffe test with the least means squares statement of the GLM procedure. The level of significance was set at $P < 0.05$.

Results

Leptin Receptor MRNA Expression in Granulosa and Theca Cells: We demonstrated the expression of leptin receptor mRNA in granulosa and theca cells from the three genotypes fed ad libitum or restricted for the S line. The expression of leptin receptor mRNA for both ovarian cells was detected in each hierarchical yellow follicle studied (F1 to F4).

Evolution of Leptin Receptor MRNA Expression with Follicular Development

The evolution of expression of leptin receptor mRNA during follicle development was investigated in both granulosa and theca cells from F4 to F1 yellow follicles using real-time RT-PCR. Since leptin receptor mRNA levels did not follow a normal distribution (skewness of 5.48 and Kurtosis of 32.88), they were log transformed. The resulting distribution was closer to the normality with skewness of 0.87 and kurtosis of 0.69. Variance analysis was performed on transformed data.

In the E line, expression of the leptin receptor mRNA decreased between F4 and F1 yellow follicles. However the high variability of the expression of leptin receptor mRNA in F4 follicles prevented the decrease from reaching statistical significance. Compared to the L line, the level of expression of the leptin receptor in the granulosa cells was lower in the E group, especially for the F4 and F3 yellow follicles but statistical significance was reached only for the F4 follicles. In the S line fed ad libitum, expression of the leptin receptor in the granulosa was dramatically up-regulated. This up-regulation was clearly evident in F4 F3 and F1 follicles. Wide variability was also observed in F4 and F3 follicles. Feed restriction of the standard hens (SR) induced a general decrease in the expression of leptin receptor mRNA. The overall level of expression of the leptin receptor mRNA measured in the SR line was similar to that observed in the L and E lines. Compared to the SA birds, feed restriction has restored the decrease in the expression of leptin receptor mRNA with follicle development.

Expression of leptin receptor mRNA in the theca cells remained stable during yellow follicle development, whatever the group of birds considered. Statistical analysis did not reveal any difference in leptin receptor mRNA expression between the 4 groups of birds.

Expression of Leptin Receptor MRNA in the Liver

In the liver, the expression of leptin receptor mRNA was up-regulated in S birds fed ad libitum. Feed restriction of S birds restored the level of expression of leptin receptor mRNA similar to that measured in the E and L birds.

Plasma Lipid and Glucose and Hormone Concentrations

At 32 weeks of age plasma leptin and insulin concentrations were found to be similar in the three genotypes. Food restriction of the standard hens did not alter plasma leptin, glucose or insulin levels. Triglyceride, cholesterol and phospholipid levels were also measured. Cholesterol and phospholipid levels were not affected by genotype or diet. On the other hand, triglyceride levels seemed to be affected by the genotype. Lower triglycerides levels were measured in Standard birds (SA, SR). However, statistical significance was reached only for the restricted SR birds.

Discussion

Several studies conducted on theca and granulosa cells have shown that leptin may have direct negative effects on ovarian steroidogenesis in various mammalian species. Leptin inhibits insulin-induced progesterone and 17β-estradiol production by isolated bovine granulosa cells and impairs the hormonally-stimulated in vitro release of 17β-estradiol by rat granulosa cells. In granulosa cells from fertile women, leptin inhibits FSH and IGF-I stimulated estradiol production Since leptin has a more potent inhibitory action of insulin-induced aromatase activity of granulosa cells from small than large follicles, it has been proposed that the numbers of leptin receptors in granulosa cells might decrease as follicles develop in order to make mature Graafian follicles less sensitive to the negative action of leptin. As shown in this study and in previous reports, the leptin receptor was expressed in the hen ovary in both granulosa and theca cells, suggesting a direct action of leptin at the level of the ovary.

It seemed that leptin might affect ovarian steroidogenesis in laying hens during fasting but the involvement of leptin on steroidogenesis during normal follicle development remained to be determined. In this study we demonstrated that the direct action of leptin on the ovary might be modified during follicle development since the level of expression of its receptor clearly decreased during maturation of yellow follicles. This decrease was particularly evident in slow growing broiler breeder hens from the "Label" genotype and from the feed-restricted standard line. Given that fast growing chickens (ad libitum-fed standard and Experimental broiler breeder hens) have the highest reproductive problems, genetic or nutritional control of the growth rate might regulate ovarian leptin receptor gene expression and improve reproductive function. Such evolution of receptor expression in the follicular hierarchy has previously been shown for the FSH receptor. FSH-stimulated steroidogenesis declined during follicle maturation and was associated with a decrease in FSH receptor numbers. Conversely, the expression of mRNA encoding the IGF-I receptor and the related efficacy of binding of IGF-I to granulosa cells increased as the follicle matured. Since leptin receptor gene expression was modified during follicle development, leptin might also be involved in regulation of the follicular hierarchy and onset of pre-ovulatory steroidogenesis, as has been proposed for gonadotrophins and growth factors including FSH and IGF-I.

Unlike mammals, progesterone in chickens is synthesised and secreted mainly by granulosa cells whereas theca cells generate estradiol. Progesterone produced by granulosa cells from mature follicles provides the positive feedback necessary to stimulate a pre-ovulatory surge of LH. IGF-I has been involved in the regulation of ovarian steroidogenesis in both mammals and birds.

IGF-I stimulates progesterone production from avian granulosa cells whereas it up-regulates estradiol from mammalian granulosa cells. Since leptin is considered to be an inhibitor of insulin and IGF-I action on steroidogenesis in mammals, leptin might have similar negative action in birds. Thus, the decrease in its receptor in the granulosa suggests that the inhibiting action of leptin would decrease during follicle development and consequently favours the stimulatory effect of gonadotrophins and IGF-I on follicular maturation.

This hypothesis is also consistent with the weaker steroidogenic response of granulosa cell culture of ad-libitum fed standard broiler breeder hens when stimulated by IGF-I compared to granulosa cell culture from feed-restricted birds. Moreover, Onagbesan et al (2004) demonstrated in an experiment similar to that performed in the present study and using the same genotypes that plasma progesterone levels were clearly affected in SA birds.

They demonstrated that plasma progesterone levels remained relatively stable between 25 and 37 weeks of age in the E, L and SA birds with a significant lower level in the SA birds (2.2 ± 0.62 ng/ml for SA birds compared to 3.9 ± 0.36 and 4.2 ± 0.54 for L and E birds respectively). In restricted standard birds, plasma progesterone levels dramatically increased and reached values (3.8 ± 0.26 ng/ml) closed to that measured in the E and L lines (Onagbesan et al., 2004, data from progesterone levels were personal communication from Dr Onagbesan, Catholic University of Leuven, Belgium).

The erratic pattern of oviposition in standard broiler breeder hens fed ad libitum has been previously demonstrated to be related to abnormal maturation of steroidogenesis, particularly in the two largest yellow follicles. Since F2 and F1 yellow follicles presented similar endocrine profiles, the pre-ovulatory surge of LH probably triggers ovulation of the two largest follicles. In this study we have shown that ad libitum feeding of broiler breeder hens dramatically up-regulated expression of the leptin receptor in the granulosa cells of yellow follicles and changed the evolution of expression of this receptor with follicle development. These results suggest a strong action of leptin on the ovaries of ad libitum fed birds.

Feed restriction reduced the level of expression of the leptin receptor and on the whole restored the evolution of expression of the receptor with follicle maturation. Since ad libitum feeding affects the hierarchical endocrine order of the follicles, as a potential inhibitor of hormonally induced avian steroidogenesis leptin represented a good candidate to explain the affects of follicular hierarchy. Up-regulation of the expression of the leptin receptor gene was also demonstrated in the liver. This up-regulation may be related to the control of lipogenesis. The liver plays a key role in lipid metabolism and lipogenesis in avian species and the standard broiler breeder hens were the fattest birds of this experiment.

Since expression of its receptor was dramatically up-regulated in SA hens, leptin probably played an important role in the increased number of large yellow follicles and abnormal follicle hierarchy. However the factors involved in regulation of the expression of the leptin receptor within the hen ovary remains to be determined. Among the plasma hormones and lipids analysed in this study only triglycerides were found to be different between strains, with a lower level in the restricted standard broiler breeder hens that were also the leanest birds. Down regulation of the expression of the leptin receptor by homologous and heterologous signals have previously been demonstrated in both mammals and chickens.

Leptin and insulin are able to down-regulate expression of the chicken leptin receptor in vitro. In the present study, plasma leptin and insulin levels were similar for each genotype and were not altered by feed restriction in the standard genotype. We have previously demonstrated that during the first 5 weeks of age, plasma leptin levels remained relatively stable in both broiler and layer chicken despite increased body weight.

However the absence of leptin levels differences may be related to the fact plasma leptin levels were measured at 32 weeks of age. SR birds were relaxed at the start of lay, switched to breeding feeding and allowed ad libitum access as the other groups of birds. We therefore suggested that leptin and insulin are probably not involved in the regulation of ovarian leptin receptor gene expression in ad libitum or feed-restricted standard broiler breeder hens.

Evidence of the regulation of expression of the leptin receptor gene in the granulosa related to follicle maturation and nutritional state strongly suggest that leptin played an important local and sequential role in the dysfunction of the follicular hierarchy observed in standard broiler breeder hens fed ad libitum. This study suggests that the level of expression of the leptin receptor regulates the action of its ligand at the level of the ovary. This provides an interesting perspective to understanding the physiological role of leptin in the ovary.

Chicken

The chicken (*Gallus gallus domesticus*) is a domesticated fowl, a subspecies of the Red Junglefowl. As one of the most common and widespread domestic animals, and with a population of more than 24 billion in 2003, there are more chickens in the world than any other species of bird. Humans keep chickens primarily as a source of food, consuming both their meat and their eggs.

The traditional poultry farming view of the domestication of the chicken is stated in *Encyclopaedia Britannica* (2007): "Humans first domesticated chickens of Indian origin for the purpose of cockfighting in Asia, Africa, and Europe. Very little formal attention was given to egg or meat production..." Recent genetic studies have pointed to multiple maternal origins in Southeast, East, and South Asia, but with the clade found in the Americas, Europe, the Middle East and Africa originating in the Indian subcontinent.

From India the domesticated fowl made its way to the Persianised kingdom of Lydia in western Asia Minor, and domestic fowl were imported to Greece by the fifth century BC. Fowl had been known in Egypt since the 18th Dynasty, with the "bird that lays every day" having come to Egypt from the land between Syria and Shinar, Babylonia, according to the annals of Tutmose III.

Terminology

In the UK, Ireland and Australia adult male chickens over the age of 12 months are primarily known as *cocks*, whereas in America and Canada they are more commonly called *roosters*. Males under a year old are *cockerels*. Castrated roosters are called *capons* (surgical and chemical castration are now illegal in some parts of the world). Females over a year old are known as *hens*, and younger females are *pullets*. In Australia and New Zealand (also sometimes in Britain), there is a generic term *chook* to describe all ages and both sexes. Babies are called *chicks*, and the meat is called *chicken*.

"Chicken" originally referred to chicks, not the species itself. The species as a whole was then called *domestic fowl*, or just *fowl*. This use of "chicken" survives in the phrase "Hen and Chickens", sometimes used as a British public house or theatre name, and to name groups of one large and many small rocks or islands in the sea.

In the Deep South of the United States chickens are also referred to by the slang term *yardbird*.

General Biology and Habitat

Chickens are omnivores. In the wild, they often scratch at the soil to search for seeds, insects and even larger animals such as lizards or young mice.

Figure 2: The adult rooster can be distinguished from the hen by his larger comb

Roosters can usually be differentiated from hens by their striking plumage of long flowing tails and shiny, pointed feathers on their necks (*hackles*) and backs (*saddle*) which are typically of brighter, bolder colours than those of females of the same species. However, in some breeds, such as the Sebright, the rooster has only slightly pointed neck feathers, the same colour as the hen's. The identification must be made by looking at the comb, or eventually from the development of spurs on the male's legs (in a few breeds and in certain hybrids the male and female chicks may be differentiated by colour). Adult chickens have a fleshy crest on their heads called a *comb or cockscomb*, and hanging flaps of skin either side under their beaks called *wattles*. Both the adult male and female have wattles and combs, but in most breeds these are more prominent in males. A *muff* or *beard* is a mutation found in several chicken breeds which causes extra feathering under the chicken's face, giving the appearance of a beard.

Domestic chickens are not capable of long distance flight, although lighter birds are generally capable of flying for short distances, such as over fences or into trees (where they would naturally roost). Chickens may occasionally fly briefly to explore their surroundings, but generally do so only to flee perceived danger.

Chickens are gregarious birds and live together in flocks. They have a communal approach to the incubation of eggs and raising of young. Individual chickens in a flock will dominate others, establishing a "pecking order", with dominant individuals having priority for food access and nesting locations. Removing hens or roosters from a flock causes a temporary disruption to this social order until a new pecking order is established. Adding hens—especially younger birds—to an existing flock can lead to violence and injury.

Hens will try to lay in nests that already contain eggs, and have been known to move eggs from neighbouring nests into their own. Some farmers use fake eggs made from plastic or stone (or golf balls) to encourage hens to lay in a particular location. The result of this behaviour is that a flock will use only a few preferred locations, rather than having a different nest for every bird.

Hens can also be extremely stubborn about always laying in the same location. It is not unknown for two (or more) hens to try to share the same nest at the same time. If the nest is small, or one of the hens is particularly determined, this may result in chickens trying to lay on top of each other.

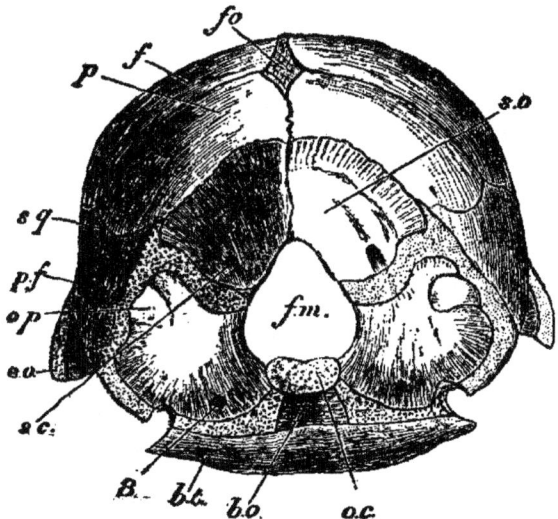

Figure 3: Skull of a chicken three weeks old. Here the opisthotic bone appears in the occipital region, as in the adult Chelonian. bo = Basi-occipital, bt = Basi-temporal, eo = Opisthotic, f = Frontal, fm = Foramen magnum, fo = Fontanella, oc = Occipital condyle, op = Opisthotic, p = Parietal, pf = Post-frontal, sc = Sinus canal in supra-occipital, so = Supra-occipital, sq = Squamosal, 8 = Exit of vagus nerve.

Roosters crowing (a loud and sometimes shrill call) is a territorial signal to other roosters. However, crowing may also result from sudden disturbances within their surroundings. Hens cluck loudly after laying an egg, and also to call their chicks. Chickens also give a low "warning call" when they think they see a predator approaching.

In 2006, scientists researching the ancestry of birds "turned on" a chicken recessive gene, *talpid2*, and found that the embryo jaws initiated formation of teeth, like those found in ancient bird fossils. John Fallon, the overseer of the project, stated that chickens have "...retained the ability to make teeth, under certain conditions...."

Food Sharing and Courting

When a rooster finds food, he may call other chickens to eat first. He does this by clucking in a high pitch as well as picking up and dropping the food. This behaviour may also be observed in mother hens to call their chicks and encourage them to eat.

To initiate courting, some roosters may dance in a circle around or near a hen ("a circle dance"), often lowering his wing which is closest to the hen. The dance triggers a response in the hen's brain, and when the hen responds to his "call", the rooster may mount the hen and proceed with the fertilization.

Breeding

Origins

The domestic chicken is descended primarily from the Red Junglefowl (*Gallus gallus*) and is scientifically classified as the same species. As such it can and does freely interbreed with populations of red jungle fowl. Recent genetic analysis has revealed that at least the gene for yellow skin was incorporated into domestic birds through hybridisation with the Grey Junglefowl (*G. sonneratii*). The traditional poultry farming view is stated in *Encyclopaedia Britannica* (2007): "Humans first domesticated chickens of Indian origin for the purpose of cockfighting in Asia, Africa, and Europe. Very little formal attention was given to egg or meat production... " In the last decade there have been a number of genetic studies. According to one study, a single domestication event occurring in the region of modern Thailand created the modern chicken with minor transitions separating the modern breeds. However, that study was later found to be based on incomplete data, and recent studies point to multiple maternal origins, with the clade found in the Americas, Europe, Middle East, and Africa, originating from the Indian subcontinent, where a large number of unique haplotypes occur.

It has been claimed (based on paleoclimatic assumptions) that chickens were domesticated in Southern China in 6000 BC. However, according to a recent study, "it is not known whether these birds made much contribution to the modern domestic fowl. Chickens from the Harappan culture of the Indus Valley (2500-2100 BC), in what today is Pakistan, may have been the main source of diffusion throughout the world."

A northern road spread chicken to the Tarim basin of central Asia, modern day Iran. The chicken reached Europe (Romania, Turkey, Greece, Ukraine) about 3000 BC. Introduction into Western Europe came far later, about the 1st millennium BC. Phoenicians spread chickens along the Mediterranean coasts, to Iberia. Breeding increased under the Roman Empire, and was reduced in the Middle Ages.

Middle East traces of chicken go back to a little earlier than 2000 BC, in Syria; chicken went southward only in the 1st millennium BC. The chicken reached Egypt for purposes of cock fighting about 1400 BC, and became widely bred only in Ptolemaic Egypt (about 300 BC). Little is known about the chicken's introduction into Africa. Three possible ways of introduction in about the early first millennium AD could have been through the Egyptian Nile Valley, the East Africa Roman-Greek or Indian trade, or from Carthage and the Berbers, across the Sahara.

The earliest known remains are from Mali, Nubia, East Coast, and South Africa and date back to the middle of the first millennium AD. Domestic chicken in the Americas before Western conquest is still an ongoing discussion, but blue-egged chicken, found only in the Americas and Asia, suggest an Asian origin for early American chickens.

A lack of data from Thailand, Russia, the Indian subcontinent, Southeast Asia and Sub-Saharan Africa makes it difficult to lay out a clear map of the spread of chickens in these areas; better description and genetic analysis of local breeds threatened by extinction may also help with research into this area.

Current

Under natural conditions, most birds lay only until a clutch is complete, and they will then incubate all the eggs. Many domestic hens will also do this–and are then said to "go broody". The broody hen will stop laying and instead will focus on the incubation of the eggs (a full clutch is usually about 12 eggs). She will "sit" or "set" on the nest, protesting or pecking in defence if disturbed or removed, and she will rarely leave the nest to eat, drink, or dust-bathe.

While brooding, the hen maintains the nest at a constant temperature and humidity, as well as turning the eggs regularly during the first part of the incubation. To stimulate broodiness, an owner may place many artificial eggs in the nest, or to stop it they may place the hen in an elevated cage with an open wire floor.

At the end of the incubation period (about 21 days), the eggs, if fertile, will hatch. Development of the egg starts only when incubation begins, so they all hatch within a day or two of each other, despite perhaps being laid over a period of two weeks or so. Before hatching, the hen can hear the chicks peeping inside the eggs, and will gently cluck to stimulate them to break out of their shells. The chick begins by "pipping"; pecking a breathing hole with its egg tooth towards the blunt end of the egg, usually on the upper side. It will then rest for some hours, absorbing the remaining egg yolk and withdrawing the blood supply from the membrane beneath the shell (used earlier for breathing through the shell). It then enlarges the hole, gradually turning round as it goes, and eventually severing the blunt end of the shell completely to make a lid. It crawls out of the remaining shell, and its wet down dries out in the warmth of the nest.

The hen will usually stay on the nest for about two days after the first egg hatches, and during this time the newly hatched chicks live off the egg yolk they absorb just before hatching. Any eggs not fertilized by a rooster will not hatch, and the hen eventually loses interest in these and leaves the nest.

After hatching, the hen fiercely guards the chicks, and will brood them when necessary to keep them warm, at first often returning to the nest at night. She leads them to food and water; she will call them to edible items, but seldom feeds them directly. She continues to care for them until they are several weeks old, when she will gradually lose interest and eventually start to lay again.

Modern egg-laying breeds rarely go broody, and those that do often stop part-way through the incubation. However, some "utility" (general purpose) breeds, such as the Cochin, Cornish and Silkie, do regularly go broody, and they make excellent mothers, not only for chicken eggs but also for those of other species—even those with much smaller or larger eggs and different incubation periods, such as quail, pheasants, turkeys or geese. Chicken eggs can also be hatched under a broody duck, with varied success.

Poultry Farming

More than 50 billion chickens are reared annually as a source of food, for both their meat and their eggs. Chickens farmed for meat are called broiler chickens, whilst those farmed for eggs are called egg-laying hens. In total, the UK alone consumes over 29 million eggs per day. Some hens can produce over 300 eggs per year. Chickens will naturally live for 6 or more years, but broiler chickens typically take less than six weeks to reach slaughter size. For laying hens, they are slaughtered after about 12 months, when the hens' productivity starts to decline, by which point they are normally infirm and have lost a significant amount of their feathers, and their life expectancy has been reduced from around 7 years to less than 2 years.

The vast majority of poultry are raised using intensive farming techniques. According to the Worldwatch Institute, 74 percent of the world's poultry meat, and 68 percent of eggs are produced this way. One alternative to intensive poultry farming is free range farming.

Friction between these two main methods has led to long term issues of ethical consumerism. Opponents of intensive farming argue that it harms the environment, creates human health risks and is inhumane. Advocates of intensive farming say that their highly efficient systems save land and food resources due to increased productivity, stating that the animals are looked after in state-of-the-art environmentally controlled facilities.

In part due to the conditions on intensive poultry farms and recent recalls of large quantities of eggs, there is a growing movement for small scale micro-flocks or 'backyard chickens'. This involves keeping small numbers of hens (usually no more than a dozen), in suburban or urban residential areas to control bugs, utilise chicken waste as fertilizer in small gardens, and of course for the high-quality eggs and meat that are produced.

Artificial Incubation

Incubation can successfully occur artificially in machines that provide the correct, controlled environment for the developing chick. The average incubation period for chickens is 21 days but may depend on the temperature and humidity in the incubator. Temperature regulation is the most critical factor for a successful hatch. Variations of more than 1 °F (1.8 °C) from the optimum temperature of 99.5 °F (37.5 °C) will reduce hatch rates.

Humidity is also important because the rate at which eggs lose water by evaporation depends on the ambient relative humidity. Evaporation can be assessed by candling, to view the size of the air sac, or by measuring weight loss. Relative humidity should be increased to around 70% in the last three days of incubation to keep the membrane around the hatching chick from drying out after the chick cracks the shell. Lower humidity is usual in the first 18 days to ensure adequate evaporation. The position of the eggs in the incubator can also influence hatch rates. For best results, eggs should be placed with the pointed ends down and turned regularly (at least three times per day) until one to three days before hatching. If the eggs aren't turned, the embryo inside may stick to the shell and may hatch with physical defects. Adequate ventilation is necessary to provide the embryo with oxygen. Older eggs require increased ventilation.

Many commercial incubators are industrial-sized with shelves holding tens of thousands of eggs at a time, with rotation of the eggs a fully automated process. Home incubators are boxes holding from half a dozen to 75 eggs; they are usually electrically powered, but in the past some were heated with an oil or paraffin lamp.

Chicken Eggs as Food

Chicken eggs are widely used in many types of dishes, both sweet and savory, including many baked goods. Eggs can be scrambled, fried, hard-boiled, soft-boiled, pickled, and poached. The albumen, or egg white, contains protein but little or no fat, and can be used in cooking separately from the yolk. Egg whites may be aerated or whipped to a light, fluffy consistency and are often used in desserts such as meringues and mousse. Ground egg shells are sometimes used as a food additive to deliver calcium. Some people prefer to just have a female, and raise it for the eggs.

Chickens as Food

The meat of the chicken, also called "chicken", is a type of poultry meat. Because of its relatively low cost, chicken is one of the most used meats in the world. Nearly all parts of the bird can be used for food, and the meat can be cooked in many different ways. Popular chicken dishes include roasted chicken, fried chicken, chicken soup, Buffalo wings, tandoori chicken, butter chicken, and chicken rice. Chicken is also a staple of many fast food restaurants.

Chickens as Pets

Chickens are sometimes kept as pets and can be tamed by hand feeding, but roosters can sometimes become aggressive and noisy, although aggression can be curbed with proper handling. Some have advised against keeping them around very young children. Certain breeds, however, such as silkies and many bantam varieties are generally docile and are often recommended as good pets around children with disabilities. Some people find chickens' behaviour entertaining and educational.

Chicken Diseases and Ailments

Chickens are susceptible to several parasites, including lice, mites, ticks, fleas, and intestinal worms, as well as other diseases. Despite the name, they are not affected by chickenpox, which is generally restricted to humans.

Some of the common diseases that affect chickens are shown below:

Name	Common Name	Caused by
Aspergillosis		fungi
Avian influenza	bird flu	virus
Histomoniasis	Blackhead disease	protozoal parasite
Botulism		toxin
Cage Layer Fatigue		mineral deficiencies, lack of exercise
Campylobacteriosis		tissue injury in the gut
Coccidiosis		parasites
Colds		virus
Crop Bound		improper feeding
Dermanyssus gallinae	Red mite	parasite
Egg bound		oversised egg
Erysipelas		bacteria
Fatty Liver Hemorrhagic Syndrome		high-energy food
Fowl Cholera		bacteria
Fowl pox		virus
Fowl Typhoid		bacteria
Gallid herpesvirus 1 or Infectious Laryngotracheitis		virus
Gapeworm	Syngamus trachea	worms
Infectious Bronchitis		virus
Infectious Bursal Disease	Gumboro	virus
Infectious Coryza		bacteria
Lymphoid leukosis		Avian leukosis virus
Marek's disease		virus
Moniliasis or Thrush	Yeast Infection fungi	
Mycoplasmas		bacteria-like organisms
Newcastle disease		virus
Necrotic Enteritis		bacteria

Omphalitis	Mushy chick disease	umbilical cord stump
Peritonitis		Infection in abdomen from egg yolk
Prolapse		
Psittacosis		bacteria
Pullorum	Salmonella	bacteria
Scaly leg		parasites
Squamous cell carcinoma		cancer
Tibial dyschondroplasia	speed growing	
Toxoplasmosis		protozoal parasite
Ulcerative Enteritis		bacteria
Ulcerative pododermatitis	Bumblefoot	bacteria

Chickens in Religion and Mythology

In Indonesia the chicken has great significance during the Hindu cremation ceremony. A chicken is considered a channel for evil spirits which may be present during the ceremony. A chicken is tethered by the leg and kept present at the ceremony for its duration to ensure that any evil spirits present during the ceremony go into the chicken and not the family members present. The chicken is then taken home and returns to its normal life.

In ancient Greece, the chicken was not normally used for sacrifices, perhaps because it was still considered an exotic animal. Because of its valour, the cock is found as an attribute of Ares, Heracles, and Athena. The alleged last words of Socrates as he died from hemlock poisoning, as recounted by Plato, were "Crito, I owe a cock to Asclepius; will you remember to pay the debt?", signifying that death was a cure for the illness of life.

The Greeks believed that even lions were afraid of cocks. Several of Aesop's Fables reference this belief.

In the New Testament, Jesus prophesied the betrayal by Peter: "Jesus answered, 'I tell you, Peter, before the rooster crows today, you will deny three times that you know me.'" (Luke 22:34) Thus it happened (Luke 22:61), and Peter cried bitterly. This made the cock a symbol for both vigilance and betrayal.

Earlier, Jesus compares himself to a mother hen when talking about Jerusalem: "O Jerusalem, Jerusalem, you who kill the prophets and stone those sent to you, how often I have longed to gather your children together, as a hen gathers her chicks under her wings, but you were not willing." (Matthew 23:37; also Luke 13:34).

In many Central European folk tales, the devil is believed to flee at the first crowing of a cock.

In traditional Jewish practice, a kosher animal is swung around the head and then slaughtered on the afternoon before Yom Kippur, the Day of Atonement, in a ritual called kapparos. A chicken or fish is typically used because it is commonly available (and small enough to hold). The sacrifice of the animal is to receive atonement, for the animal symbolically takes on all the person's sins in kapparos. The meat is then donated to the poor. A woman brings a hen for the ceremony, while a man brings a rooster. Although not actually a sacrifice in the biblical sense, the death of the animal reminds the penitent sinner that his or her life is in God's hands.

The Talmud speaks of learning "courtesy towards one's mate" from the rooster (Eruvin 100b). This might refer to the fact that when a rooster finds something good to eat, he calls his hens to eat first.

The chicken is one of the Zodiac symbols of the Chinese calendar. Also in Chinese religion, a cooked chicken as a religious offering is usually limited to ancestor veneration and worship of village deities. Vegetarian deities such as the Buddha are not one of the recipients of such offerings. Under some observations, an offering of chicken is presented with "serious" prayer (while roasted pork is offered during a joyous celebration). In Confucian Chinese Weddings, a chicken can be used as a substitute for one who is seriously ill or not available (e.g. sudden death) to attend the ceremony. A red silk scarf is placed on the chicken's head and a close relative of the absent bride/groom holds the chicken so the ceremony may proceed. However, this practice is rare today.

A cockatrice was supposed to have been born from an egg laid by a rooster, as well as killed by a Rooster's call.

Chickens in History

An early domestication of chickens in Southeast Asia is probable, since the word for domestic chicken (**manuk*) is part of the reconstructed Proto-Austronesian language. Chickens, together with dogs and pigs, were the domestic animals of the Lapita culture, the first Neolithic culture of Oceania.

The first pictures of chickens in Europe are found on Corinthian pottery of the 7th century BC. The poet Cratinus (mid-5th century BC, according to the later Greek author Athenaeus) calls the chicken "the Persian alarm". In Aristophanes's comedy *The Birds* (414 BC) a chicken is called "the Median bird", which points to an introduction from the East. Pictures of chickens are found on Greek red figure and black-figure pottery.

In ancient Greece, chickens were still rare and were a rather prestigious food for symposia. Delos seems to have been a centre of chicken breeding.

The Romans used chickens for oracles, both when flying ("ex avibus", Augury) and when feeding ("auspicium ex tripudiis", Alectryomancy). The hen ("gallina") gave a favourable omen ("auspicium ratum"), when appearing from the left (Cic.,de Div. ii.26), like the crow and the owl.

For the oracle "ex tripudiis" according to Cicero (Cic. de Div. ii.34), any bird could be used, but normally only chickens ("pulli") were consulted. The chickens were cared for by the pullarius, who opened their cage and fed them pulses or a special kind of soft cake when an augury was needed. If the chickens stayed in their cage, made noises ("occinerent"), beat their wings or flew away, the omen was bad; if they ate greedily, the omen was good.

In 249 BC, the Roman general Publius Claudius Pulcher had his chickens thrown overboard when they refused to feed before the battle of Drepana, saying "If they won't eat, perhaps they will drink." He promptly lost the battle against the Carthaginians and 93 Roman ships were sunk. Back in Rome, he was tried for impiety and heavily fined.

In 161 BC, a law was passed in Rome that forbade the consumption of fattened chickens. It was renewed a number of times, but does not seem to have been successful. Fattening chickens with bread soaked in milk was thought to give especially delicious results. The Roman gourmet Apicius offers 17 recipes for chicken, mainly boiled chicken with a sauce. All parts of the animal are used: the recipes include the stomach, liver, testicles and even the pygostyle (the fatty "tail" of the chicken where the tail feathers attach).

The Roman author Columella gives advice on chicken breeding in his eighth book of his treatise on agriculture. He identifies Tanagrian,

Rhodic, Chalkidic and Median (commonly misidentified as Melian) breeds, which have an impressive appearance, a quarrelsome nature and were used for cockfighting by the Greeks. For farming, native (Roman) chickens are to be preferred, or a cross between native hens and Greek cocks. Dwarf chickens are nice to watch because of their size but have no other advantages.

Per Columella, the ideal flock consists of 200 birds, which can be supervised by one person if someone is watching for stray animals. White chickens should be avoided as they are not very fertile and are easily caught by eagles or goshawks. One cock should be kept for five hens. In the case of Rhodian and Median cocks that are very heavy and therefore not much inclined to sex, only three hens are kept per cock. The hens of heavy fowls are not much inclined to brood; therefore their eggs are best hatched by normal hens. A hen can hatch no more than 15-23 eggs, depending on the time of year, and supervise no more than 30 hatchlings. Eggs that are long and pointed give more male, rounded eggs mainly female hatchlings.

Per Columella, chicken coops should face southeast and lie adjacent to the kitchen, as smoke is beneficial for the animals. Coops should consist of three rooms and possess a hearth. Dry dust or ash should be provided for dust-baths.

According to Columella, chicken should be fed on barley groats, small chick-peas, millet and wheat bran, if they are cheap. Wheat itself should be avoided as it is harmful to the birds. Boiled ryegrass (*Lollium* sp.) and the leaves and seeds of alfalfa (*Medicago sativa* L.) can be used as well. Grape marc can be used, but only when the hens stop laying eggs, that is, about the middle of November; otherwise eggs are small and few. When feeding grape marc, it should be supplemented with some bran. Hens start to lay eggs after the winter solstice, in warm places around the first of January, in colder areas in the middle of February. Parboiled barley increases their fertility; this should be mixed with alfalfa leaves and seeds, or vetches or millet if alfalfa is not at hand. Free-ranging chickens should receive two cups of barley daily.

Columella advises farmers to slaughter hens that are older than three years, because they no longer produce sufficient eggs.

Capons were produced by burning out their spurs with a hot iron. The wound was treated with potter's chalk.

For the use of poultry and eggs in the kitchens of ancient Rome see Roman eating and drinking.

Chickens were spread by Polynesian seafarers and reached Easter Island in the 12th century AD, where they were the only domestic animal, with the possible exception of the Polynesian Rat (*Rattus exulans*). They were housed in extremely solid chicken coops built from stone.

Chickens in South America

An unusual variety of chicken that has its origins in South America is the araucana, bred in southern Chile by Mapuche people. Araucanas, some of which are tailless and some of which have tufts of feathers around their ears, lay blue-green eggs. It has long been suggested that they predate the arrival of European chickens brought by the Spanish and are evidence of pre-Columbian trans-Pacific contacts between Asian or Pacific Oceanic peoples, particularly the Polynesians and South America.

In 2007, an international team of researchers reported the results of analysis of chicken bones found on the Arauco Peninsula in south central Chile. Radiocarbon dating suggested that the chickens were Pre-Columbian, and DNA analysis showed that they were related to pre-historic populations of chickens in Polynesia. These results appeared to confirm that the chickens came from Polynesia and that there were transpacific contacts between Polynesia and South America before Columbus's arrival in the Americas.

However, a later report looking at the same specimens concluded:

> *A published, apparently pre-Columbian, Chilean specimen and six pre-European Polynesian specimens also cluster with the same European/Indian subcontinental/Southeast Asian sequences, providing no support for a Polynesian introduction of chickens to South America. In contrast, sequences from two archaeological sites on Easter Island group with an uncommon haplogroup from Indonesia, Japan, and China and may represent a genetic signature of an early Polynesian dispersal. Modelling of the potential marine carbon contribution to the Chilean archaeological specimen casts further doubt on claims for pre-Columbian chickens, and definitive proof will require*

further analyses of ancient DNA sequences and radiocarbon and stable isotope data from archaeological excavations within both Chile and Polynesia.

Drinking and Feeding Systems

Feeding Systems

There are several different systems available for delivery and distribution of feed to broilers. Since feed constitutes the major share of total production cost, wastage should be an important consideration in the choice of system.

There are three major systems available:

- Automatic pan feeders: 1 pan per 65 birds; 33 cm pan diameter.
- Chain feeders: 2.5 cm per bird; 80 birds per metre of track.
- Round, hanging tube feeders: 65 birds per tube; 38 cm diameter base.

Automatic pan feeding systems have become the industry standard due to advantages of low feed wastage, ease of height adjustment, preservation of pellet quality, and reliability. As a number of different pan feeder designs are available, feeder heights should be set according to manufacturers' recommendations. Distance between the feeder lines should be not more than 2.5 metres. This ensures that all birds have adequate access to feed. Level of feed within the feeder should be adjusted to a height that minimises wastage. If possible, the feed supply system should be allowed to empty at least once a day. This eliminates the presence of stale food and therefore reduces the risk of contamination and the growth of micro-organisms.

Drinking Systems

It is essential that fresh water is available to the broiler flock at all times and that it is free of contamination. The drinking systems chosen must be capable of delivering the water efficiently to all birds with the minimum of spillage. To ensure that the flock is receiving sufficient water, each day, the ratio of water to feed consumed should be monitored.

When the ratio of water volume (ml or l) to feed weight (g or kg) remains close to 1.8:1 (1.6:1 for nipple drinkers), only then can it be assumed that the birds are consuming sufficient water. Birds will drink more water at high ambient temperatures. Water requirement increases by approximately 6.5% per degree as temperature exceeds 21°C. Water consumption will vary with feed consumption.

Nipple Drinkers

Nipple systems provide water with lower levels of bacterial contamination than conventional open systems. They have become the standard in modern broiler production.

General recommendations for the management of nipple systems are:

- 12 birds per nipple. This should be reduced to 9-10 per nipple for birds weighing 2.75 kg or more.
- Nipple height should be monitored daily and adjusted as appropriate. At day old, nipples should be placed at chick eye level. From day 2 onward, while drinking, the back of the chick should form an angle of 45° with the floor.
- Litter, under and around the drinker lines, should be level to allow all birds to have equal access to water.
- Drinker lines should be level to a avoid spillage.
- Individual nipples should be checked regularly to confirm that access is available to birds through 360° (i.e. from all directions). Faulty nipples will reduce birds' access to drinking water. Nipples should be activated and checked by hand before placement to ensure all nipples are working.
- Water pressure should be set according to manufacturers' specifications.
- Nipple lines should be flushed and sanitised weekly.

Bell Drinkers

- When whole house brooding is practiced, a minimum of 6 bell drinkers should be provided per 1000 chicks.
- Drinkers should be distributed evenly throughout the house so that no broiler is more than 2 m from water.
- As a guide to level, water should be 0.6 cm below the top of the drinker until 7-10 days and there should be 0.6 cm of water in the base of the drinker from 10 days onwards.
- The height at which the bell drinkers are suspended should be checked and adjusted daily, so that the lip of the bell is level with the broilers' backs from 7 days onwards.

Agritech

Comparative evaluation of feed conservation in fibreglass and metal silos during summer and winter time

Agritech srl has entrusted the Faculty of Agriculture of the Università Cattolica del Sacro Cuore, in cooperation with Cerzoo, the Research Center for Zootechnics and Environment of Piacenza (Italy) with a comparative study on the performance of some own-manufactured fibreglass silos and other silos in galvanised metal, produced by Chore Time. The study covers a period of time going from July 21st 2008 to February 2nd 2009.

Aim of the Study

The aim of the study was the comparative analysis of the conservation of bulk feed in mealy form stored in fibreglass and galvanised metal silos during summer and winter time.

Compared Samples

1. Storage silos

    The study was carried out on a total of 6 silos of 6 ton capacity each, supplied by Agritech, 3 of which in fibreglass (VTR) and 3 in metal (MET). The silos were installed in pairs (VTR-MET) in the same environmental conditions with regard to the exposure to the sun.

2. Stored feed

    The feed selected for the test was commercial compound feed in mealy form, taken from the same production batch, and stored in the silos in the same quantity.

    The silos were only loaded by 2/3 of their real capacity in order to reproduce the normal conditions of use in a standard farm, where silos are progressively emptied. To put in evidence eventual effects connected with non-ideal storage and environmental conditions typical of summer, some vegetal oil was added to the main feed as lipidic integration.

3. Period

    The study was carried out from July 2008 to February 2009.

4. Surveys

    During the 4-month summer test following surveys were carried out:

a) *Temperature (T°):* Using Min. and Max. thermometers, the values of T° MAX., T° MIN. and T° INSTANT in the external environment, the air temperature inside the silos (the empty volume between the cover of the silos and the surface of the feed) and the temperature of the feed (with a thermometer being placed in the first 10 cm of the bulk) were measured on alternate days.

b) The amount of peroxides released over 20 days, based on feed samples taken from the superior and inferior part of each silo.

During the 3-month winter test following surveys were carried out:

c) *Temperature (T°):* Using Min. and Max. thermometers, the values of T° MAX., T° MIN. and T° INSTANT in the external environment, the air temperature inside the silos (the empty volume between the cover of the silos and the surface of the feed) and the temperature of the feed (with a thermometer being placed within the first 10 cm of the bulk) were measured on alternate days.

5. *Results:* As far as the summer period is concerned, it was noticed that:

a) There are remarkable differences between the temperature of the air inside the silos and that of the stored feed in relation to the building material of the silos. These differences, that arise from the majority of the surveys, are statistically significant, and they are indicated in the charts by some asterisks corresponding to the date of the survey. As one can observe, the large amount of significant marks proves that fibreglass silos can stand thermal fluctuations and control both the air and the feed temperature better than metal silos.

In fibreglass silos (VTR), instant, minimum and maximum temperatures resulted to be better than in metal silos (MET) in the majority of the surveys.

Particularly, the Authors observed higher MAX. T° of the feed in the upper part of the bulk and of the air inside metal silos, as reported in chart No. 3 Delta T° between VTR and MET.

As reported, these differences reach up to over 8°C in the air inside silos, with registered MAX. T° over 45°C in the metal silos. The temperature of feed in the upper part of silos has reached temperatures between 35°C and 45°C in the metal silos, while in the fibreglass silos the highest temperature never exceeded 35°C (this peak was only reached in three surveys). The temperature deltas (Ä) comparatively registered in the temperatures of the feed stored in VTR or MET silos confirm the better performance of fibreglass silos, with a difference in favour of fibreglass up to 7°C.

b) Regarding the amount of released peroxides registered in the upper and lower part of the feed bulk, the fibreglass silos show a lower rate of peroxides and consequently a lower oxidation of the lipids contained in feed than metal silos (...).

Regarding the winter period (November 2008 – February 2009), the Authors had following results:

a) "Max. temperature of the air inside the silos: all temperature data registered in fibreglass silos are statistically inferior to those registered in metal silos. The average difference measured during the 15 test-weeks is about - 106,32%. These values in fibreglass silos tend to match with those of the environmental temperature in the "hottest" weeks, while, similarly to what happens with the minimum temperatures, the lowest ("coldest") values tend to be inferior to those measured in the outside environment".

b) "Max. temperature of the feed inside silos: in the majority of the surveys, the MAX. T° of the feed stored inside fibreglass silos show values that are much inferior to those registered in metal silos.

"Finally, the results obtained in the second part of the test, which integrate and complete the results of the summer test, prove that also in the winter months the temperature registered in the air and feed inside silos is the parameter which is subject to the most significant variations, and that it largely depends on the building material of the silos. In fibreglass silos, both the temperature of the inside air and the temperature of the stored feed

bulk are averagely inferior to the temperatures registered in metal silos".

This e-mail address is being protected from spambots. You need JavaScript enabled to view it

Big Dutchman

ReproMatic and FluxxBreeder – new feeding system especially for broiler breeders.

ReproMatic is a feeding system developed by Big Dutchman to ideally meet the particular requirements of broiler breeder management. Only this system allows all birds to receive feed immediately and simultaneously. It combines the advantages of chain and pan feeding. The rugged feed chain is used as conveying system. The feed channel with chain allows for a high filling level and consequently a very high conveying capacity.

FluxxBreeder is Big Dutchman's newly developed feed pan that can be used in rearing for day-old to death production and in broiler breeder production.

The main aim in rearing is for all hens of a flock to reach laying maturity at the same time. A uniform flock can only develop, if all birds have sufficient space to feed. Moreover, pans must be filled simultaneously at the same speed and to the same level to allow all birds of a flock to receive the same amount of feed during restricted feeding.

Features of FluxxBreeder in the rearing phase:

ReproMatic-pullets: The same view with pullets.

16 "true" feeding spaces, which means 60 % more birds per running metre of feeding system as compared to a linear trough;

- 360° flooding mechanism that ensures a high feed level in the pan, especially in the first days of rearing;
- spin-n-lock system allows for simple, one-handed adjustment of the feed level;
- flat pan for an ideal start of day old chicks, a good distribution of feed and reduced feed losses;
- the elevated feed channel and rotatable pan provide the birds with enough freedom of movement;

- eight wings on the outer cylinder of the pan prevent feed losses as lateral feed spillage is not possible;
- the integrated volume reducer allows for small feed rations, thus ensuring fast and simultaneous filling of all pans;
- after the birds have been moved out, the pans can easily be cleaned with a high-pressure cleaner, for drying the pan bottom can simply be opened;
- excess cleaning water or disinfectants can easily drain off through additional holes in the pan bottom;
- the conveying system consists of a feed channel equipped with the Challenger feed chain, thus allowing the transport of large amounts of feed with a high conveying capacity (2 t/h);
- smooth feed saving lip prevents bruises and feed losses;
- sliding shut-off to close individual pans;
- ideal illumination of the pan due to openings in the pan top.

Amacs is Big Dutchman's sophisticated and modern management system for the control of the entire feed supply but also for climate, water supply, bird weighing and lighting. Amacs allows to control a low-maintenance batch. Since the feed lines can be filled when they are suspended there is no need for a large hopper.

Amacs has a modular design and can be used for small and large houses alike, as it can be adapted to the individual situation on a farm. Amacs allows for ongoing data collection, real-time control and monitoring of individual barns or entire farm complexes – all this from virtually any location in the world.

Chore-Time

Brock Europe offers poultry drinking and feeding systems.

Chore-Time Europe B.V. offers drinker and feeder options for commercial poultry producers.

Chore-Time's popular RELIA-FLOW® Nipple Drinker offers producers a reliable flow rate consistent with the way birds actually drink. This university-tested drinking system features robust, precision-machined, stainless steel parts in the flow-control area of the valve which resist wear and retain their shape for long life and consistent, reliable flow. The system includes a one-piece valve for easy field replacement in retrofit applications. Chore-Time also offers a basic drinker model with its STEADI-FLOW® Nipple Drinker.

For turkey producers, Chore-Time offers the ADVANTI-FLOW® Poult Drinker. The ADVANTI-FLOW Drinker features pockets in the drinker's disc that hold attractive beads of water. The pockets and the disc's scalloped edge work together to control the direction of water flow. The disc also provides leverage for easier triggering of the nipple drinker by young birds. Dual catch cups with rounded edges for bird comfort maximise water consumption while helping keep floors dry.

Chore-Time recently introduced a new drinking system regulator which is easier for users to install, manage and maintain. Designed for long, reliable service, the regulator requires less pressure to seal than other regulators, resulting in less wear on the sealing mechanism. Simple-to-use, top-mounted knobs activate the "regulate" or "flush" modes for each drinker line, while a bottom knob is used to adjust the operating water column at each regulator.

Also available from Chore-Time is a new, large-diameter, folding stand tube which folds to protect both the tube and the ceiling during house clean out. The rigid PVC stand tube requires no spring for support, making the water column easier to see and helping the tube to stay cleaner. The large-volume pipe removes more air from the watering system than smaller diameter tubes.

Chore-Time offers poultry producers convenient control of the water pressure levels in all nipple drinker lines in the house with its updated PDS™ Controls. With Chore-Time's PDS (Pneumatic Drinking System) Controls, users can change the pressure in all lines in the house or flush all lines from one remote location.

One of Chore-Time's most popular feeding systems is its Model C2® PLUS Pan Feeding System. The feeder is available in a standard model, as well as models with a shallow pan to help with starting birds or a model with extended fins to enhance feed flow for harder-flowing feed types. The chick-friendly 14-spoke grill design includes feed-saving features to help maximise feed conversion throughout the growing cycle.

Chore-Time's also offers its unique REVOLUTION® Poultry Feeder. Unlike conventional feeder pans with flood windows, the REVOLUTION Feeder includes a Rotary Feed Gate which provides full control of both flood and final feed levels without needing to depend on pan height or gravity to operate. On the floor or in the air, the Rotary Gate gives the REVOLUTION Feeder complete control of both flood and final feed levels.

Chore-Time Europe B.V. is an affiliate of CTB, Inc. Based in Milford, Indiana (U.S.A.), CTB, Inc. is a leading global designer, manufacturer and marketer of systems and solutions for the poultry, pig, egg production, and grain industries. Its products and services are "Helping to Feed a Hungry World®" through improved efficiency and air quality management in the care of poultry and livestock as well as in grain storage, handling, conditioning and drying.

Founded in 1952, CTB has been dedicated to "Leadership Through Innovation®" throughout its history. The company operates from multiple locations in various countries around the world and serves its customers through a worldwide network of independent dealers and distributors.

Codaf

Figure 4: The Feeding Pans from Gio Series by Codaf

Angiolino Daffi, who is the founder of the Italian poultry equipment company Codaf, has been working in the poultry sector since 1962. Codaf is specialised in production of automatic poultry feeding system.

The Italian company is the only one all over the world in providing Gio, a patented feeding pan which can be used with conventional supply tubes without having to resort to particular modifications.

Innovative range of GIO feeding-pans series confirms its setting out characteristic to have an easier access to the feed.

The innovative design of the pan, without grille, added to the special edge profile, allows the birds to easily reach the feed even during the first days, thus achieving better results. The patented self-adjusting feed level system of the feeder pans is very simple to operate, requiring no adjustment through grow-out. Manual regulation of the feed level is extremely simple, by raising and slightly turning the yellow regulator. The simplicity enables correct feed level at every age of the flock, thus obtaining record feed conversion rates.

With an easy and rapid cleaning at the end of the cycle GIO feeding-pans produce a better uniformity of the reared animals rending these pans always more requested and spread in the world.

Euro Agro Products

The Nipplemakers of the World

Euro Agro Products BVBA is located in Waregem, Belgium and it's specialised in manufacturing nipple drinkers for poultry, pig and cattle industry. More than 40 years of experience, competitive price and top quality have enabled EAP to be well known worldwide as "The Nipplemakers of the World".

Because of many years of experience, Euro Agro Products BVBA is always ready to assist its customers to ensure good results and a steady cooperation.

In the past years the assortment of drinkers has expanded to suit everyone's request and needs.

EAP is now launching its new type of nipple drinkers made from stainless steel. The nipple drinkers have a 360° range and deliver a standard flow of 40-60ml/min or 60-80ml/min. The nipple

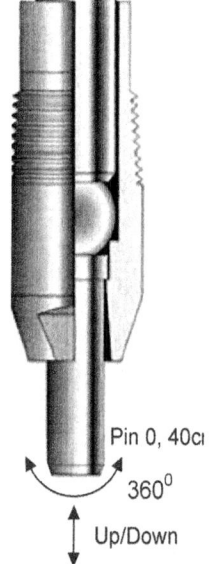

Pin 0, 40ci

360°

Up/Down

drinkers are made from stainless steel, material code 304, which means they have a long lifespan.

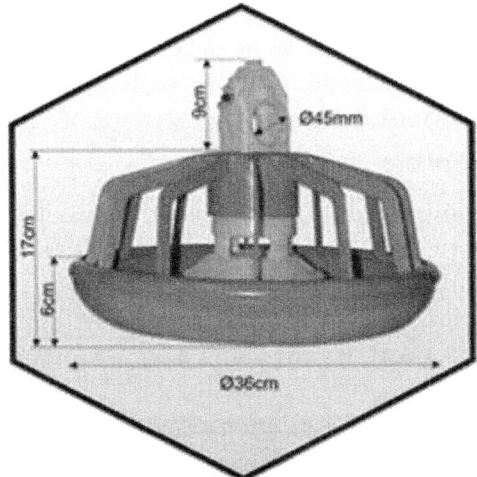

In the last two years EAP has been busy developing a new type of feeder pan. The result is a solid 800 gram first class PVC feeder pan. This combined with EAP's first class nipple drinkers give poultry operators a strong set, ideal for the birds.

Euro Agro Products BVBA is also specialised in complete poultry house projects involving watering, feeding, fogging, lighting, heating and cooling systems. For Turn Key projects the Company is able to deliver and install the Flex auger and silo as required.

Roxell

Full Range of Roxell Poultry Feeding and Drinking Concepts

Figure 5: HaiKoo: Easy access to the feed.

HaiKoo™ for broilers has a low pan edge for easy access by day old birds. The large feeding surface offers optimal eating comfort to the young birds. The 360 degrees assures fast flooding of the pans. Practice shows that HaiKoo guarantees an excellent start. The exclusive grill, the high feed saver rim, the anti-spill wings and the feed trough deep in the pan avoid feed waste.

Uniform Layer Pullets

Poolaï™ is developed for the rearing of pullets. The low pan edge allows day-olds to spot the feed within seconds. The 360° flooding keeps the pan full of feed. The oval shaped pan is equipped with a high feed saver rim and has an incorporated feed trough in the bottom. A special pan lock prevents the pans from swinging to avoid spillage.

Poola

Figure 6: Uniform egg production.

LaiCa™ was developed for layers in production and for layer parent stock.

Figure 7: Roxell's layer range

A uniquely designed pan in cascade shape prevents selective pecking. Special wings on the cone stop the hens from grubbing in the feed while a wide feed saver rim prevents spilling.

12 wide grill openings allow easy access to the feed. LaïCa is applicable in different housing systems: floor feeding (free range or bio), aviary systems and enriched cages.

Excellent Hatchability

KiXoo™ has a refined inner grill to apply easy and efficient separate sex feeding. The grill adjusts to anywhere between 40 and 50 mm in mm increments by turning a handy adjuster knob on top of the pan lid. The grill also adapts in height simply by clicking the inner grill in a 90, 80 or 70 mm position.

Both elements prevent male breeders having access to the hens' feed, no matter the age or breed.

High Uniformity of Young Hens

ViToo™ optimises the feeding chances for day old broiler breeders when arriving in the house. The feed windows allow an excellent filing of the pans to guarantee that the chicks find the feed immediately. The low pan rim and the wide grill openings give day olds comfortable access to their feed. After a couple of days already birds start eating from the outside of the pan, lining up neatly around it.

An easy shut-off slide on the pans allows partial brooding.

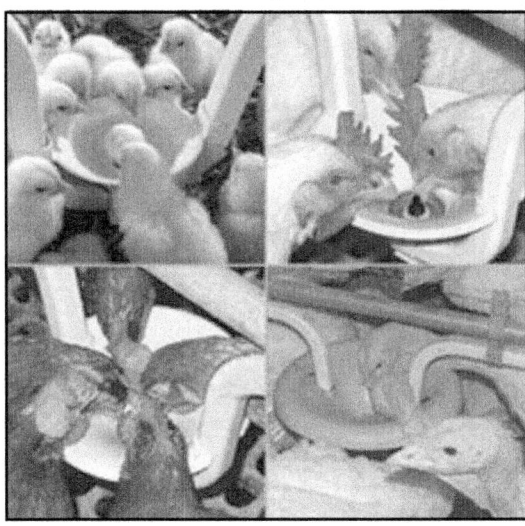

Figure 8: Full feeder range for broiler breeders

Drinking Systems for All Types of Poultry

Roxell offers a full range of drinking systems adaptable to any type of poultry.

SPARKcup™ is a unique cup drinker with different standard, medium or large sized drinker cups. It offers broilers, broiler breeders, layers, turkeys, etc. excellent access to plenty of clean, fresh water.

Water spillage is practically excluded which guarantees dry litter. This leads to excellent hygiene and improves the house climate. SPARKcup is easy to flush, during and between rounds, which prevents sediment in the pipes and around the nipples.

Roxell's SPARKnipple™ drinking system includes solid drinking nipples and small or big drip cups. A wide choice of nipples allows matching the drinking system perfectly to the different poultry types.

Ska

Ska Feeding Systems: In the ever-increasing world of different feeding systems, it is vital to choose something that has been specifically designed for its purpose, a solution that offers solid construction, simplicity of operation, optimal feed access and minimal manual intervention.

Ska, with 55 years of experience in poultry floor equipment, offers the widest range of automatic feeders:

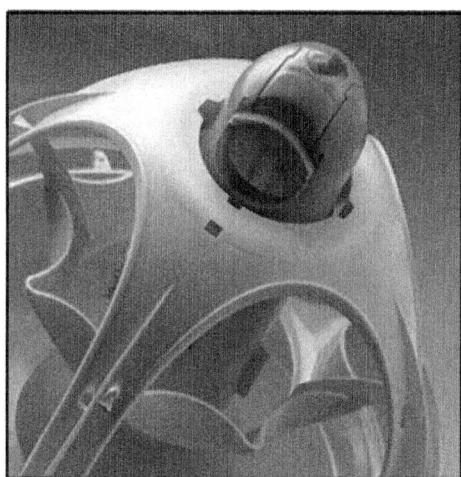

Figure 9: Lyra

- *FLATLINE:* Chain feeder for breeders, pullets and commercial layers, with several combinations, as it can be on legs, installed on the slats or, by means of special winches, it can be completely suspended and winched daily. It is available with different single and double speed motors, coupled with several troughs designs and various grill versions. It can be easily customised or upgraded, thanks to accessories can be added at any time.

- *VENUS:* Breeder pan feeder, highly reliable, with homogeneous feed distribution and easy to operate. It enables 14 females to feed simultaneously and prevents cocks to enter the pan. Suitable for all-in, all-out houses too.

- *ATOS:* Cockerel pan feeder. Supported by a strong and stable galvanised structure, with the widest possible feed access. A pan designed and engineered specifically for male breeders, it can accommodate 10 to 12 cockerels.

- *LYRA:* Multipurpose pan feeder with several combinations of grids and pans to guarantee excellent results with broilers, turkeys, ducks, guinea-fowls, commercial layers and cockerels. It eliminates feed wastage thank to a double rim, and requires only one simple centralised feed level adjustment whenever feeding day-old chicks or adult birds. Available also with the new drop tube LYRA T for seamless feeding of turkey, from day 1 to final day.

- *LEO:* Broiler pan feeding system, which easily allows farmer to use it for day-old chicks without any supplementary feeders, making sure they stay outside the pan. It drastically reduces feed waste, and requires only single, centralised adjustments throughout the cycle.

- *GAMMA:* Broiler pan feeding system, simple and robust, easy feed level adjustment and high quality materials, the ideal solution for a tight budget without compromising on quality.

- *SUPERPITO:* A feeder to cover all and every request of modern turkey rearing, with its strong galvanised wires and two different pans for growing and production, easy to replace and assuring no spillage, no matter the size of birds. Fully centralised feed adjustment.

- *MIRA:* The only quail feeder on the market, which avoids wastages, accidental trapping and injuries to these small birds.

VDL

Feeding Systems: Feeding systems of VDL Agrotech are characterised by:

- Design for optimal technical performance for each application
- The feeding system design should be simple but effective
- Easy to install, user friendly and durable
- First class materials
- The system must be easy to clean.

Figure 10: Chain Feeding System of VDL Agrotech

For more than 40 years the VDL chain-feeding systems is a well-known and widely used system all over the world. It is the chain feeding system that can be used for any type of the poultry.

This high quality product is available in numerous configurations, 1 up to 4 circuits, various speeds up to 36mtr/min. System can be either supported or suspended. All components are manufactured by VDL. High quality and durability are guaranteed.

Pan Feeding Systems

Classic pan was first introduced 25 years ago but still a popular model sold worldwide.

The clients specially value its strong but simple construction combined with excellent technical results and minimum feed spillage.

The Bridex pan with unique male excluder grill is highly suitable for breeder operations. The adjustable height of the opening in the male excluder grill allows adjustment to suit the required settings for each breed. The Bridex pan incorporates a feed quantity restrictor allowing distribution of small feed portions. The Bridex pan is characterised by simple design, durability, low maintenance and excellent performance. Bridex can be supplied for both straight line (45 mm Ø) and circular systems (60 mm Ø).

Figure 11: Pan Feeding Systems

Fittra pan makes the feed accessible for birds from day 1. Its open design prevents birds from getting caught inside the feed pan. Despite only 57 mm high profile of the feed pan is the feed spillage reduced to a minimum. Due to unique design of the pan also ideal for heavy birds. The 36 cm diameter of the feed pan offers additional feeding space.

Flextra Pan: This unique product is designed to serve professional rearing farms for rearing of breeder pullets, to facilitate optimal feed intake from first until last day. The unique double tube system for uniform feed distribution takes care that all feed pans are filled simultaneously, contributing to high uniformity of the system. The Flextra control system enables feed distribution within minimum of time, preventing pans from running empty during feeding.

Flex-system 2009 is the new designed modular flex auger system of VDL Agrotech. Different modules and variety of integrating combinations are the strong points of the Flex system 2009:

- The VDL Agrotech extension intake hopper allows installing of extra augers between the siloboot (single or double) and the

"Flex 2009" feed intake hoppers and is available for all Flex sizes (Ø 60, 75, 90 and 125).

- The VDL Agrotech module silo is a modular build-up storage system that can be used for intermediate storage of feed mixtures in the house, feed kitchen or service room. The basic set with storage bin of 500 litres total (390 + 110ltr), can be extended with a maximum of 3 extensions of 800 litres each.
- VDL Agrotech designed also an universal weigher support fram.

5

The Problem of Floor Eggs

Floor eggs are a serious issue in modern broiler breeder production for a number of reasons. These eggs are more likely to be contaminated, resulting in a number of associated problems. Contaminated eggs have a reduced hatchability and increase the likelihood of omphalitis in chicks that do hatch. The egg is an ideal culture medium for eggs and, when incubated, the contaminating bacteria can grow prolifically, resulting in an egg that will not only not hatch but may actually explode, contaminating other eggs in the incubator. On the farm, a high prevalence of floor eggs means significantly more labour to collect them and, of course, lost income due to those that are discarded or do not hatch. It is in the best interest of hatching egg producers to minimise floor eggs and, if a problem is identified, to do everything possible to correct the situation.

Nesting Behaviour in Hens

Before exploring the causes of floor eggs, it is interesting to review the normal nesting behaviour of chickens. There is a series of behavioural actions that ancestral chickens and other related birds follow when nesting and brooding eggs. Modern commercial birds retain many of these traits, but to variable degrees.

Hens will naturally seek a secluded area to nest. Ideally, it will be in a hidden location, protected from the view of potential predators. They will approach the nest indirectly and cautiously, always on the lookout for predators. The construction of the nest is generally not complicated, particularly if the location is favourable. When the hen is ready for nesting, a male may lead her to the nest area. This behaviour may be manifested as the gathering into corners that is often observed in modern breeder males.

Once on the nest, hens remain motionless to minimise any attraction by predators.

Variations in the nesting behaviour of modern hens that laid floor eggs compared with those that laid eggs in nest boxes have been observed experimentally. In a study done by Cooper and Appleby (1996), a group of 20 hens was housed individually and observed for nesting behaviour and site selection for egg laying.

Six of the 20 birds were found to be persistent floor egg layers. Those that were floor-layers performed more nest seeking behaviour, less nest building behaviour, and less sitting prior to laying eggs. Floor-layers appeared to find the nest boxes unattractive, either because they had lower nesting motivation or they perceived the nest box as an inappropriate site for nesting. Floor eggs declined with age.

Causes of Floor Eggs in Modern Broiler Breeder Flocks

In modern broiler breeder birds there are a number of factors, ranging from genetics through environment to early rearing, that can contribute to an increase in floor eggs. It has been suggested that the behavioural characteristics for nesting are retained by modern chickens but, because of genetic intervention over time, certain aspects of that behaviour are modified. Such modifications are different among various strains. Thus, in the Ross bird, for example, there are differences in the expected number of floor eggs among the various strains. The Ross 308 can be expected to produce 3 to 6% floor eggs, the 508, 5-7%, and the 708, 1-2%.

An appropriate number of nest boxes must be available. The maximum ratio is 5.5 hens per nest box. Soiled nest boxes will also discourage occupation by the hens, effectively reducing the number of sites available. Too few nest boxes will cause some hens to seek out less appropriate locations. It is not unusual, for example, for some hens to find the slight shade under water lines or feed lines as a favourable location.

Aggressive cockerel activity can contribute to nest box refusal. Separate male feed lines provided in the scratch area will encourage the cockerels to spend more time away from the slats. Hens will then be more likely to find the nest boxes as a favourable location to lay their eggs. A higher than optimum cockerel to hen ratio can also cause some aggressive behaviour.

A number of environmental conditions may encourage hens to lay eggs on the floor. As described previously, hens will look for a spot that is apparently secure and hidden in which to lay their eggs. So, if there are areas of cover or distinct shadows, some birds may find that area preferable to the nest box.

For example, hens will often use the shaded area under ramps that have been provided for access to the slats as a nesting area. Irregular lighting or bulbs that are burned out can cause distinct shadows being cast by equipment, and the birds can perceive such areas as a suitable nesting location.

Irregular airflow may also contribute to inappropriate nesting behaviour. If air movement results in drafts in the nest boxes, the hens may choose areas in the scratch that are more comfortable. This may be seen as specific locations in which eggs chronically accumulate.

Training pullets to access perches during the rearing period is also an important contributor to success in appropriate nesting behaviour. Research has shown that, when birds are not allowed access to perches before 6 weeks of age, they will not gain the ability or inclination to use the perches during the lay period. It is generally recommended that the flock be allowed access to training perches by 4 weeks of age.

Correcting Inappropriate Nesting Behaviour

Once birds in a flock have started to lay floor eggs, it can be difficult to correct the problem. Generally the number of floor eggs will go down with age, although the numbers may still be higher than desired. Any action taken will be directed at making the nest boxes more attractive to the hens and making the inappropriate areas less attractive.

The design of the modern boxes is generally good and quite acceptable to the hens. Nest boxes must be clean and dry, however, and have comfortable substrate on the floor. Make sure they are cleaned of manure regularly. They should have a perch area at the front to allow the hens to check out the box before entering.

The environment in and around the box must also be suitable. Make sure that the interior of the boxes is shaded or protected from bright light. They should also be sheltered from drafts. Such conditions will encourage hens to use the nest boxes as nest areas.

Temperature should be maintained between 68° F and 72° F consistently. It is perhaps even more important that the temperature be consistent throughout the house. This will prevent pockets of comfortable airspaces that hens may find in which to nest. Similarly, air quality should be maintained with good ventilation so air is fresh throughout the barn.

If floor eggs are a problem, it is important that the barn be walked several times daily. Floor eggs should be picked up frequently and any floor nests – that is depressions in the litter that have been made by hens attempting to nest – should be filled in. Any hens attempting to nest in the scratch should be disturbed and encouraged up to the slat area. To minimise any effect of aggressive male behaviour, make sure the cockerel to hen ratio is correct. Remove any sub optimal males – those with bad legs, those that are small, or those that are not mature.

A high incidence of floor eggs can be a serious problem for the hatchery as well as the hatching egg producer. Steps taken early in the life of a flock, including proper care and feeding, assuring good uniformity, and training to perches can go a long way in minimising floor eggs. Other important factors to be addressed include good nest design and location, good environmental conditions, including light, and proper male to female ratio. Once a floor egg problem has been identified, it can be difficult to correct, but attending to management basics may help to reduce the impact.

Achieving and Maintaining Fertility in Broiler Breeders

Infertility...Male or Female Problem?
- It could be both...
- However, if an egg is laid...the "potential" fertility is always there.
- Males, may be or may become "impotent".

Infertility...
- In females mostly due to: Over-feeding and obesity
- In males mostly due to: Over-feeding and under-feeding.

Infertility...
- If a hen produces an egg... infertility may be due simply to absence of   semen in oviduct...(most common case)

- This in turn may be due to:
- Mating frequency or mating success.

That's why males are more critical but often overlooked.

Fertilization

A fertile egg:

1. Must have gone through a fusion of male and female gametes.
2. Must have a properly developed embryo at oviposition.

Fertilization... When / How It Happens ?

- Sperm: egg interaction:
- After copulation, small % of spermatozoa enter the sperm storage tubules (utero-vaginal junction).
- Spermatozoa are released around 30% per day, to the infundibulum during 10 minutes around ovulation fertilization ?
- During the 10 minutes ovum moves down and secretions involve it, preventing any more sperms to attach.

Fertilization... When / How ?

- Only the pronucleus of one spermatozoa is allowed to fuse with the female gamete at the center of germinal disc (4 – 5 hours)
- Number of holes in perivitelline membrane is important.
- How many spermatozoa are required? A total of approx. 1000 around de ova ...with 6 holes around the germinal disc.

Fertilization ...When / How ?

- Surprisingly, this process is incredibly efficient...
- Under most conditions fertility level reaches over 90% ...
- However... the % of infertile eggs can be of significant economic importance.
- Highest fertility is the result of good rearing and adult period management of males and females.

Perivitelline Membrane (IPVL) Penetration:

- Spermatozoa "tunnel" into the germinal disc area using proteolytic enzymes.
- Then, oviductal secretion form the OPVL to prevent to excess sperm activity in the area.

- Fertilization depends on good quality semen (sperm count, motility).

- Supplied by a well developed male with optimum testicular growth (puberty, early adulthood).

Feed Restriction Significantly Alters Lipogenic Gene Expression in Broiler Breeder Chickens

The modern commercial broiler is the product of intensive selection over many generations for rapid growth and enhanced muscle mass. Selection for these economically important traits has been accompanied by an increase in voluntary feed intake, resulting in birds that do not adequately regulate feed intake to achieve energy balance. Thus, broiler chickens are prone to obesity resulting from hyperphagia when given free access to feed.

Although unrestricted feeding of these birds with typical starter and breeder rations would ensure an adequate nutrient supply to support growth and development, most commercial broiler feeding programmes utilise varying levels of feed restriction as a management tool to regulate body weight, improve egg production and promote flock uniformity during the rearing and breeding phases of production. There have been many reports comparing ad libitum vs. restricted feed intake and changes in feed composition on various aspects of welfare, growth, body composition and egg production in broiler breeders. Because egg production causes major changes in the metabolism of lipids by breeder hens to meet the demands of yolk formation, it is important to understand the underlying genetic mechanisms governing lipid metabolism, specifically as they are affected by nutritional status.

Birds have the ability to store large quantities of excess energy (in the form of triglycerides) in liver, adipose tissue and in yolk of developing oocytes. Lipogenesis (i.e., the conversion of glucose to triglycerides) takes place primarily in the liver of birds and involves a series of linked, enzyme-catalysed reactions including glycolysis, the citric acid cycle and fatty acid synthesis. Hepatic lipogenesis is subject to both nutritional and hormonal control and this metabolic process is highly responsive to changes in the diet. Adipose tissue serves primarily as a storage site for lipid with little lipogenesis occurring in this tissue. Differential lipogenic capacity of liver vs. adipose tissue in birds is a function of the expression of a key transcription factor, sterol regulatory element binding protein-1 (SREBP-1). The gene for SREBP-1 is highly expressed in the liver, but to a much lesser extent in adipose tissue. Moreover, the expression of a number of lipogenic enzyme genes such as fatty acid synthase (FAS), malic enzyme (ME), acetyl CoA carboxylase (ACC), ATP citrate lyase (ACL) and steroyl CoA (Ä9) desaturase 1 (SCD1) is directly influenced by SREBP-1. Lipid accumulation by adipose tissue and the developing oocyte depends on plasma lipid that is derived from hepatic lipogenesis and lipid absorbed from the diet.

In fact, it has been estimated that 80–85% of the fatty acids present in broiler adipose tissue triglyceride stores were derived from hepatic lipogenesis or from the diet via intestinal absorption. Therefore, hepatic lipogenesis and the export of lipid are crucial steps linked to adipose tissue lipid accretion, as well as to oocyte growth and maturation (i.e., yolk formation) in egg-laying hens.

The nutritional state of the bird, as determined by the amount and composition of feed consumed, dramatically affects hepatic lipid metabolism. Variations in nutrient intake and status are communicated to the liver and other internal organs by alterations in the plasma levels of hormones and metabolites (e.g., glucose, free fatty acids) that respond acutely to dietary changes. To determine the metabolic consequences of feed restriction on body composition and egg production in broiler breeder pullets, we investigated changes in the levels of key metabolic hormones and the expression of selected hepatic lipogenic genes during the pullet-to-breeder transition period.

Materials and Methods

Stock and Management: Cobb 500 broiler breeder pullets were fed a controlled amount of feed to maintain optimum body weight according to Cobb Breeder Management Guide specifications until they reached 21 wk of age. At this time, the birds were placed in individual laying cages and half were switched to unrestricted feeding (ad libitum) with Breeder I feed (11.82 MJ/kg metabolisable energy; 16 g crude protein/100 g feed); the remaining birds were fed a restricted amount of the same ration according to Cobb guidelines. At 22 wk, all birds were photostimulated and maintained throughout a laying cycle ending at 36 wk. Photostimulation consisted of changing the photoperiod from 8 h light:16 h dark to 12 h light:12 h dark. Body weights were determined and samples of plasma, liver and abdominal fat pad were collected at the following times: *1*) 1 d before photostimulation (prelight) at 22 wk, *2*) at first egg (<"24 wk) and *3*) through peak egg production (plateau, 36 wk). Total egg production was monitored through 36 wk for some birds (plateau groups). Tissue samples were snap frozen in liquid nitrogen at the time of their collection and stored at -80°C before analysis. Plasma was stored frozen (-20°C) before analysis. All procedures followed established protocols approved by institutional animal care and use committees.

Hormone and Enzyme Assays: Specific immunoassay techniques were used to determine plasma levels of insulin, glucagon (kit, Linco Research, St. Charles, MO), T_3 (18) and 17ß-estradiol (kit, Diagnostics Products, Los Angeles, CA). Plasma samples were treated according to kit protocols or as described previously. Cytosolic ME (EC 1.1.1.40) activity in liver tissue samples was determined and expressed as μmol of oxidised or reduced NADP/(min · g liver) at 30°C as described previously.

Gene Expression Analyses: Total RNA was isolated using the TRIzol reagent according to the manufacturer's protocol (Invitrogen/ Life Technologies, Carlsbad, CA). RNA integrity was assessed via agarose gel electrophoresis and RNA concentration and purity were determined spectrophotometrically using A_{260} and A_{280} measurements. Reverse transcription (RT) reactions (20 μL) consisted of 1 μg total RNA, 50 U SuperScript II reverse transcriptase (Invitrogen/Life Technologies), 40 U of an RNAse inhibitor (Invitrogen/Life Technologies), 0.5 mmol/L dNTP, and 100 ng random hexamer primers. Polymerase chain reaction (PCR) was performed in 25 μL containing 20 mmol/L Tris-HCl, pH 8.4, 50 mmol/L KCl, 1.0 μL of the RT reaction, 1.0 U of Platinum Taq DNA polymerase (Hot Start, Invitrogen/Life Technologies), 0.2 mmol/L dNTP, 2.0 mmol/L Mg^{2+} (Invitrogen/Life Technologies), 10 pmol each of the gene specific primers and 10 pmol each of the primers specific for ß-actin. Thermal cycling parameters were as follows: 1 cycle 94°C for 2 min, followed by 30–35 cycles, 94°C for 30 s, 58–60°C for 30 s, 72°C for 1 min with a final extension at 72°C for 8 min.

Capillary Electrophoresis with Laser-Induced Fluorescence Detection (CE/LIF): Aliquots (2 μL) of RT-PCR samples were diluted 1:100 with deionised water before CE/LIF. A detailed description and validation of the CE/LIF technique used in this study for quantitative analysis of gene expression was reported previously (20). Briefly, a P/ACE MDQ CE instrument (Beckman Coulter, Fullerton, CA) equipped with an argon ion LIF detector was used. Capillaries were 75 μm i.d. x 32 cm μSIL-DNA (Agilent Technologies, Folsom, CA). EnhanCE dye (Beckman Coulter,) was added to the DNA separation buffer (Sigma, St. Louis, MO) to a final concentration of 0.5 g/L. Samples were loaded by electrokinetic injection at 3.5 kV for 5 s and run in reverse polarity at 8.1 kV for 5 min. Integrated peak area for the PCR products separated by CE was calculated using P/ACE MDQ software (Beckman Coulter,).

Quantitation of Gene Expression: The level of gene expression was determined as the ratio of integrated peak area for each individual gene PCR product relative to that of the coamplified ß-actin internal standard. Values are presented as the mean ± SEM of 5 individual determinations.

Statistical Analysis: Data were analysed by two-way ANOVA using the general linear models (GLM) procedure of SAS software (SAS Institute, Cary, NC). Significant differences among individual group means were determined with Duncan's multiple range test option of the GLM procedure of SAS software. Pearson correlation coefficients for the interrelationship of selected variables were determined using the PROC CORR procedure of SAS software. Significance was set at $P < 0.05$. Linear regression analysis was used to relate changes in gene expression values (peak area ratios) with enzyme activity measurements for ME.

Results

Feed restriction during egg production resulted in significantly ($P < 0.05$) lower body and abdominal fat pad weights compared with unrestricted feeding. Also, egg production was higher with a significantly ($P < 0.05$) lower incidence of abnormal eggs in the restricted compared with the ad libitum birds. Feed restriction produced significant ($P < 0.05$) effects on circulating levels of key metabolic hormones before the onset of egg production. Pullets that had been fed a restricted amount of feed for 21 wk before being switched to an ad libitum feed intake exhibited dramatic changes in the levels of insulin, glucagon and T_3. Circulating insulin and T_3 levels were significantly ($P < 0.05$) higher and glucagon levels were significantly ($P < 0.05$) lower in the ad libitum compared with the restricted birds before photostimulation (prelight). Before photostimulation, the molar ratio of insulin to glucagon was 20.20 ± 2.85 vs. 1.44 ± 0.11 for the ad libitum and restricted birds, respectively. When the birds began to lay eggs, this ratio increased significantly ($P < 0.05$) in the restricted group to 9.72 ± 1.11, whereas it declined significantly ($P < 0.05$) in the ad libitum group to 6.99 ± 1.02 during this time. As egg production progressed to plateau, both groups exhibited declines in this ratio.

Limiting feed intake during egg production produced significant ($P < 0.05$) effects on the expression of hepatic genes involved in lipid metabolism, particularly those genes regulating lipogenesis including ME, ACL, ACC, FAS and SCD1. In addition, expression of the transcription factor SREBP-1 was influenced in a similar manner by feed restriction.

For the ad libitum group in general, expression of each of these genes relative to ß-actin was highest before the onset of egg production (prelight) and declined as the birds came into and maintained egg production. The feed-restricted breeders exhibited significantly ($P < 0.05$) lower lipogenic gene expression levels before photostimulation compared to their ad libitum counterparts. In the restricted group, peak expression levels were delayed until after photostimulation through the onset of egg production (first egg). This was followed by a rapid decline during the plateau phase of the laying cycle.

Hepatic expression levels of apolipoprotein (apo)VLDL-II, apoB and fatty acid binding protein (FABP), genes that code for proteins involved in lipid transport, were also determined. On the basis of expression profiles in both the ad libitum and restricted birds, it was clear that each of these genes was regulated differently from those involved in lipogenesis. This most likely reflects the involvement of estrogen in the regulation of gene transcription. In general, expression levels relative to ß-actin increased significantly ($P < 0.05$) for both ad libitum and restricted groups as the birds came into egg production and remained significantly ($P < 0.05$) elevated during the laying cycle compared with prelight levels. The exact opposite was observed for the apoA1 gene. Expression levels declined throughout the period of egg production in both groups.

In contrast to the lipid transport genes, lipoprotein lipase (LPL) gene expression in adipose tissue exhibited a pattern similar to the lipogenic enzyme genes. This could indicate some coordination between genes regulating lipid production in liver and LPL that mediates lipid uptake by extrahepatic tissues. Feed restriction during early egg production significantly ($P < 0.05$) increased metallothionein (MT) gene expression over that of ad libitum birds before photostimulation, and again at the plateau phase of egg production. Expression of the isocitrate dehydrogenase (ICDH) gene in liver followed a pattern similar to that observed for MT. Both MT and ICDH gene expression profiles were opposite to those observed for the lipogenic enzyme genes. The up-regulation of hepatic MT and ICDH genes was positively correlated with elevated circulating glucagon levels ($r^2 = 0.87$ and 0.86, respectively) and may signal the presence of stress in the feed-restricted birds. No significant ($P > 0.05$) effects were noted for the expression of the leptin receptor (LR) gene, although there was a trend in both groups towards an increased expression throughout egg production. Although the differences among groups at each of the experimental sampling periods were small and not significant ($P > 0.05$), this trend indicated a possible role for estrogen in the regulation of LR gene expression.

Significant ($P < 0.05$) correlations were found among the expression levels of various lipogenic genes. Some of the highest correlations were found among genes encoding hepatic enzymes linked in the lipogenic pathway such as FAS, ACC and SCD1. SREBP-1 gene expression was positively correlated with all of the lipogenic enzyme genes studied. This undoubtedly reflects the role of this key transcription factor in coordinating hepatic lipogenesis. In addition, adipose tissue LPL expression was positively correlated with all of the major lipogenic enzyme genes, possibly indicating a functional coordination in the expression of this gene affecting lipid deposition in adipose tissue with those genes regulating hepatic lipid production.

Significant ($P < 0.05$) correlations were also identified among circulating levels of key metabolic hormones and the level of expression of the genes involved in lipogenesis. Insulin and T_3 were positively correlated with hepatic lipogenic genes, whereas glucagon exhibited a negative correlation with each of these genes.

Circulating 17ß-estradiol levels were significantly ($P < 0.05$) correlated with the level of expression of genes involved in lipid transport. FABP, apoVLDL-II and apoB were positively correlated with 17ß-estradiol, whereas apoA1 was negatively correlated. Significant ($P < 0.05$) negative correlations were also found between the level of expression of the ApoA1 gene and those for FABP, apoVLDL-II and apoB. Expression levels of FABP, apoVLDL-II and apoB were positively correlated with each other, perhaps indicating a functional interrelationship of these genes in hepatic lipid transfer, especially during egg production.

An excellent linear relationship ($y = 0.087323x - 0.259947$, $r^2 = 0.9999$) between hepatic cytosolic ME enzymatic activity and expression of the ME gene was observed for the ad libitum birds at each of the sampling periods and for the restricted birds before photostimulation. However, this relationship did not hold true for the restricted birds during egg production (first egg and plateau phases), suggesting the existence of additional post-transcriptional and/or post-translational mechanisms (e.g., mRNA turnover or allosteric control) for regulating enzyme activity level within the livers of these birds during this period. In contrast, changes in ME activity in the ad libitum group of birds appeared to be regulated exclusively at the level of gene transcription.

Discussion

This study compared the effects of feed restriction on the expression of selected lipogenic genes in broiler breeders during the pullet-to-breeder transition period. Unrestricted feeding of breeders during egg production led to increased body and abdominal fat pad weights and lowered egg production efficiency compared with birds subjected to restricted feeding.

It has been suggested that excessive body weight gain, resulting from overfeeding of female birds during the breeding phase of production, accelerates ovarian follicular maturation such that more ovulations occur than the oviduct can effectively process. This leads to an increase in the production of defective or nonsettable eggs. The higher incidence of abnormal (nonsettable) eggs produced by the ad libitum group in this study is consistent with that suggestion. Because the average chicken egg yolk contains <"4 g of triglycerides, sustained egg production is an energy-intensive process, requiring a large increase in the supply of lipid (triglycerides) to support the demands of new yolk formation.

This increased supply of lipid can originate either from hepatic lipogenesis or from intestinal absorption of dietary lipid. Because typical poultry rations are relatively low in fat (<10 g/100 g feed), the liver synthesises a major portion of lipid destined for deposit in adipose tissue and the ovary of the laying hen. In breeder hens, excessive accumulation of lipid in adipose tissue reduces feed efficiency and diverts necessary energy supply away from egg production.

Nutritional (energy) status and the subsequent responses of key plasma metabolic hormones (insulin, glucagon and T_3) are important factors that determine the level of hepatic lipogenesis in birds. Lipogenesis is dependent on glucose metabolism to provide the acetyl CoA necessary to initiate and sustain de novo fatty acid synthesis. Carbohydrate (glucose) availability, and thus lipogenic activity, would be expected to be higher in ad libitum compared with restricted birds. This was in fact the case as indicated by significantly ($P < 0.05$) higher expression of hepatic SREBP-1, ME, ACL, ACC, FAS and SCD1 genes in ad libitum compared with restricted birds just before photostimulation. Feed restriction also significantly ($P < 0.05$) affected circulating levels of insulin, glucagon and T_3. Both insulin and T_3 induce a number of genes coding for lipogenic enzymes including ME, ACC and FAS, as well as the key transcription factor that coordinates the majority of lipogenic enzyme gene expression, SREBP-1. Glucagon, on the other hand, specifically inhibits the expression of the SREBP-1 gene and lipogenic enzyme genes such as ACC and SCD1.

The surge in lipogenic gene expression in ad libitum birds before photostimulation undoubtedly reflects increased feed consumption occurring in this group after the earlier period of feed restriction (up to 21 wk of age), and it tended to diminish as the birds came into egg production. This surge was accompanied by elevated plasma levels of insulin and T_3 and reduced glucagon levels in ad libitum birds compared with those maintained on restricted feeding. These responses are analogous to what occurs in response to starvation followed by refeeding, which provides additional glucose to the liver and alters circulating levels of insulin, glucagon and T_3 to enhance lipogenic activity. In the restricted group, the up-regulation of lipogenic genes was delayed until after photostimulation (first egg).

This may have been the result of an increased allotment of feed offered to the birds during this time or the effect of increased circulating estrogen level in response to photostimulation, which enhances hepatic lipogenic activity in laying hens to meet the demands for egg yolk formation. Specific changes in plasma hormones (i.e., increased insulin and T_3 with lowered glucagon) produced a physiologic state, promoting lipogenesis in the restricted birds at the time of first egg compared with the prelight period.

At the plateau phase of egg production, expression levels of ACC, FAS, SCD1 and FABP genes were significantly ($P < 0.05$) higher in the livers of ad libitum birds compared with their restricted counterparts. Because the major mode of regulation of hepatic lipogenesis by nutritional status is at the level of gene transcription, this could signal a higher rate of lipogenesis in the ad libitum birds at this time. It could also have contributed to the increased abdominal fat pad size observed in ad libitum compared with restricted birds. It is interesting to note that the reaction catalysed by ACC constitutes the rate-limiting step in the fatty acid synthetic pathway and that, in our study, each of the genes significantly ($P < 0.05$) up-regulated by unrestricted feeding was at or below the ACC step in this pathway. Perhaps coordinate expression of these functionally interrelated genes was established in response to the level of feeding.

An indication of this might be found in the high degree of correlation among expression levels of these hepatic lipogenic genes. Hillgartner et al. previously reported that dietary control of FAS and ME gene transcription is coordinated with that of ACC in chicken liver, and they further suggested that common control mechanisms are involved in the nutritional regulation of hepatic lipogenic enzyme genes in chickens. The coordination of lipogenic gene transcription in response to nutritional status appears to be tissue specific and most likely involves unique *cis*-acting sequences and *trans*-acting factors present in the liver.

Because mRNA levels do not always correlate directly with the amount of functional protein produced within cells, determining the functional relationship between gene expression measurements (i.e., mRNA levels) and enzyme activity is one way to gauge the relevance of the expression data to actual physiologic/biochemical effects. We provided one example of this, showing a direct linear relationship between hepatic gene expression values and measurements of cytosolic enzyme activity for ME. In ad libitum birds, this direct relationship held at each of the three phases of production monitored. However, it did not apply to restricted birds during egg production (first egg and plateau). This suggested the involvement of other undefined factors/processes regulating the level of ME gene transcription and/ or enzyme activity. Short-term adaptive changes in ACC enzyme activity can be achieved by covalent modification of the protein (i.e., phosphorylation) and allosteric control (by citrate) in addition to long-term transcriptional regulation mediated by insulin, glucose, T_3 and glucagon. Similar short-term mechanisms have not been reported previously for the regulation of ME activity. Although the discrepancies between ME expression and enzyme activity data in the restricted group at first egg and plateau production phases of this study remain unexplained, they may indicate additional post-transcriptional regulation, perhaps involving mRNA stability or translational efficiency.

Two cytosolic, NADP(+)-dependent hepatic enzymes (ME and ICDH) that provide reducing equivalents in the form of NADPH were investigated in this study. Malic enzyme has previously been suggested to provide the majority of reducing equivalents required by FAS for fatty acid biosynthesis in the liver of birds. On the basis of the differential patterns of gene expression observed for ME and ICDH, one could speculate that ICDH does not play a similar role.

Instead, ICDH may perform a completely different function, such as supplying NADPH for the regeneration of glutathione and other systems involved with intracellular defence against oxidative damage. Similarly, MT has also been suggested to play a role in energy metabolism by acting as an intracellular antioxidant. It is important to point out that the pattern of gene expression observed for both MT and ICDH in this study was quite similar. Also, the expression of both genes was highly correlated with circulating levels of glucagon, suggesting a role for glucagon in the regulation of these two genes. Elevated plasma glucagon is indicative of the presence of stress accompanied by reduced lipogenesis, increased lipolysis and increased potential for the generation of reactive oxygen species as by-products of fatty acid oxidation. However, the actual role(s) that ICDH and/ or MT might play, if any, in the maintenance of cellular redox state and energy homeostasis during stress in broiler breeders remains to be elucidated.

In birds, the accumulation of lipid in extrahepatic tissues results to a large extent from the combined effects of hepatic lipogenesis and lipoprotein production. Therefore, plasma triglyceride levels are dependent on the level of lipogenesis that takes place in the liver and on the action of systems involved in the packaging and export of triglycerides in the form of VLDL particles. In birds, specific mechanisms exist to partition energy between the ovary (developing yolk) and adipose tissue stores during periods of active egg production. With the onset of egg production, a shift in the type of VLDL produced by the liver occurs. In response to increased circulating levels of estrogen, the liver redirects the VLDL assembly process towards the production of a new and smaller subclass of lipoprotein particles. This new VLDL particle, designated "$VLDL_y$" for yolk-targeted, contains large amounts of apoVLDL-II in addition to apoB. The presence of apoVLDL-II appears to specifically inhibit the action of LPL, thus making $VLDL_y$ unavailable to tissues via LPL hydrolysis. Instead, the developing yolk follicle, via a receptor-mediated process that specifically recognises the apoB component, assimilates the intact $VLDL_y$ complex. The transfer of triglycerides from plasma VLDL into adipose and other tissues such as skeletal and cardiac muscle involves the action of LPL, which hydrolyses triglycerides to fatty acids and glycerol. The fatty acids are then taken up and reesterified to form new triglyceride deposits. Although it has been suggested that adipose tissue LPL is somewhat resistant to nutritional or hormonal changes, little is known about LPL gene regulation in birds. The fact that abdominal fat pad size increased in both groups of birds in this study during periods of active egg production indicates that lipid continues to be deposited into adipose tissue in addition to yolk. Moreover, it has been reported that overfeeding laying hens can negatively affect egg production, alter plasma lipoprotein profiles and, in extreme cases, lead to an excessive accumulation of lipid in the liver, giving rise to fatty liver hemorrhagic syndrome.

In conclusion, a better understanding of the mechanisms governing the partitioning of lipid stores between adipose tissue and ovarian follicles (yolk) is required to develop strategies to effectively control energy metabolism in female broiler breeders. Moreover, studying changes in lipogenic gene expression in response to restricted vs. ad libitum feed intake should provide useful information for evaluating energetic efficiency after changes in the feeding regimen. The effect of such events on the partitioning of energy stores among different extrahepatic tissues in broiler breeders warrants further study.

Managing Small Poultry Flocks

Care of Poultry to Increase Productivity

A little increase in the care provided to a small poultry flock will improve livability, growth rate and egg production. The goal is to improve productivity with only a small increase in the cost for feed or the time spent to care for the flock. Protecting young birds from rain and predators and giving feed supplement for 2 to 4 weeks will often double or triple the number of chicks that survive from a hatch, compared to allowing them to scavenge with the hen as soon as they are hatched. Vaccination against Newcastle disease and other infections will also improve the health of the flock. Hens prefer to lay in a hidden nest. If nests are provided in a darkened area raised from the ground, more eggs are recovered.

Poultry are kept under a variety of husbandry systems:

- *Scavenging:* The flock is made up of chickens of various ages of both sexes. The flock is allowed to forage freely in the village, forest or fields and along roadsides.

  — The lowest level of husbandry is when no care is given and when chickens obtain feed and water from the area where they live. They roost in trees at night. Hens usually lay only one or two clutches of eggs each year in hidden nests.

  — Night shelters provided. Small houses, coops or cages on stilts or raised off the ground are provided for night-time protection from predators and the weather. These pens should be well ventilated but openings should be small enough to exclude predators and vermin. Wire mesh may be used for this purpose. The shelter should be at least 60 cm high and roosts should be provided in the shelter.

  — Enclosed yard and night shelter. The night shelter would be as in No. (II). Kitchen and garden waste and water could be provided in the yard which would be designed to exclude goats, pigs and other animals but would allow the poultry to roam freely.

- *Backyard Flocks with Night Shelter:* Yarded or backyard flocks are confined to a large or small fenced area where they receive more care and shelter. These may be flocks of mixed age and sex but might also be groups of chickens raised for meat (broilers) or hens kept for egg production. The owner may allow other birds to scavenge outside the yard.

  — *Partial Scavenging:* The fenced area may enclose fruit or vegetable gardens or a field where the confined birds scavenge for some of their feed. Water is supplied. Fresh feed in the form of garden or kitchen waste, cultivated or wild plants, seeds, industrial feed waste, etc. may be given free choice, with the residue removed each day to be fed to pigs, goats, etc. Grain or other dried feed with or without concentrate or commercial feed may also be provided.

For laying hens a darkened, raised nesting area with nests containing clean nesting material should be available. Nests can be in the fenced area or in the shelter. The entrance to the nests should be through a covered passageway or tunnel in front of the nests. If the nests are in the shelter the back of the nests should be covered but may be made to open from the outside of the shelter for egg collection. Hens will lay many more eggs if artificial light is available in the morning and evening to give 15 to 16 hours of light each day. If most of the extra light is given in morning (starting at 0300 hr.) many of the chickens will lay before noon. Extra light should also be given in the evening when days start to shorten so that total hours of light do not become less.

— *All Feed Provided:* Flocks may be confined to a large or small yard with shade trees and a shelter. No feed is available other than that provided as fresh or dry feed.

• *Pens or Cages:* Chickens may be confined to pens or cages for their entire life, or in the case of layers when they approach laying age. Owners may hatch and raise their own replacements or buy eggs from their neighbours to hatch under hens or in an incubator. They can buy chicks from a hatchery or chicks or older birds from their neighbours.

— *Cages:* Satisfactory cages can be hand made from small sticks or sticks split in half or quarter, or from split bamboo strips 1 cm wide. There should be a 1 to 2 cm space between floor slats and floor slats should be outside (round side) up. The floor slats are supported on large (5 to 8 cm dia) straight poles spaced 10 to 20 cm apart carried by a wooden frame or posts set in the ground. Spaces between slats may be wider on the sides and top. The sides, top and door can be tied or fastened to form a grid or lattice or woven. Water and feed troughs can be made from boards, or if available

5 to 12 cm diameter bamboo poles. Spillage is reduced if only the top 1/3 of the bamboo is cut out (rather than having it split in half). The feeder and waterer is usually placed outside the cage. The opening to the feed and water must be large enough for birds to put their heads through easily and should be horizontal bars. Fresh forage and other fresh feed can be placed on a tray inside or outside the cage (residue removed daily). Housed chickens receiving roughage or fibrous material require insoluble grit (small hard stones) to grind the fibre. For layers the floor of the cage can be sloped in the direction of the floor slats to allow eggs to roll out of the cage to a holding area at the front or back of the cage (as with commercial wire cages). Since hens prefer a flat floor only a slight slope is necessary for eggs to roll out.

Cages can be small for 1 to 3 hens or large for up to 8 to 10 hens. Large cages can have roosts and hide-away nests like small shelters. Cages should be 0.5 to 1.5 m off the ground. The cages do not have to be in a building. They can be outside. Outside cages must be covered to provide shelter from rain and sun. If the cover is metal there should be a wood or cardboard ceiling under the metal, with air space between, to stop radiant heat reaching the chickens. Cages may be in a single row or back to back under one roof. Shelter from wind may also be necessary. Water must always be available. Homemade cages are good for as few as 3 to 5 hens or for several hundred birds.

— *Traditional Buildings:* Round or rectangular clay, or clay and straw, brick walled houses with pole and wood, metal or thatched roof are suitable for small poultry flocks in many hot climates. There should be a ventilation and light opening of 0.5 to 1 m between the walls and roof and an opening at the centre or peak for ventilation. These openings should be screened to prevent the entry of wild birds and predators. The floor should be raised to prevent flooding. Dried grass, straw or other material can be used to cover the floor. This type of building can be used to house laying hens or broilers and can also be used as a night shelter for chickens that forage or are confined to a yard.

III. *Commercial or Modified Commercial Poultry Sheds:*
Commercial poultry sheds should have a concrete floor for
chickens raised on the floor or a raised wire or slat floor
where the droppings fall through to the ground. The pen
should be located in a well-drained area to prevent flooding.
Depending on the climate the sides may be wire or the
lower part may be wood, plastic, concrete or other material
if protection from cold or wind is important. Adequate
ventilation is important. If the roof is metal there should
be a ceiling to prevent radiant heat from striking the
chickens. The roof may slope in one direction or have a
peak in the centre. The centre peak may be left open for
ventilation and protected from rain by a raised cover over
the opening or one side of the roof raised above and
protruding over the other side.

Commercial broiler sheds have all plastic or partial plastic
sides that can be lowered on hot days or raised for brooding
young chicks to provide extra heat from the brooders and
at night. The sides can also be raised in windy or wet
weather.

Laying hens may be kept in commercial wire cages in open
sheds, or in sheds with wire sides to exclude wild birds and
predators.

Selection of Breeders to Improve Poultry Production

Many government agriculture agencies have breeding programmes
for local breeds of chickens to improve the meat and egg production
of village and small-holder poultry flocks. They use selection of the
larger, rapidly growing and higher egg producing hens and their male
offspring to provide breeding stock or replacements to people who
want to keep small flocks of poultry on a scavenging, semi-confinement
or total confinement (in pens or cages) system. Poultry owners can
use similar methods to improve their own backyard flocks. Remove
inferior males. Select the better hens for special care so that more
eggs are produced, and protect their chicks to ensure a higher survival
rate. This could also be done at a village or cooperative level so that
better breeding stock is available in the village.

The Model Farm Concept: With community, district, country
or international assistance a local person is selected as a model for

better production skills. Several people could be selected for different crops or types of livestock production, or a variety of models could be on one farm in the village. Better methods of food crop production and harvest would be practiced on this farm. Information on better, different, or disease resistant seeds and trees would be available. Breeding stocks of poultry, pigs and sheep would be on display and available for loan or purchase. Refrigeration for keeping poultry and animal vaccines, medicine and equipment would be located here or close by. Someone would be trained to use the equipment and vaccines.

Minimising the Effect of Hot Climates

Birds do not have Sweat Glands: They cannot cool themselves by water evaporation from the skin as people can. When they are hot they pant and cool themselves by evaporation from the throat and respiratory tract. Poultry that are growing rapidly or laying eggs have a high metabolic rate and produce heat that they must lose. Chickens die from hyperthermia when their body temperature gets too high. High humidity increases the danger. Broilers and broiler breeders are most at risk, but layers and other poultry also may die because they become too hot. Egg production and growth rate are reduced when chickens are too hot. There are many ways to reduce the effect of heat on poultry.

Building location and construction:

- *Trees:* With small buildings or cages natural shade from large trees protects the chickens from the sun. The shade is most important in the afternoon and evening. Trees also have a cooling effect because of evaporation from their leaves.

- *Wind and Sun:* Large commercial style buildings should be placed where they take advantage of wind. If possible the pens should be built sideways to the prevailing wind so wind blows through the pen from side to side. It is best if the pen is placed so the sun rises and sets over the ends of the buildings (or there is a wide roof overhang) so morning and afternoon sun does not shine on the chickens. This is very important for hens in cages that cannot move away from the sun's rays. The roof slope should be low or slope in only one direction to improve air movement. A reverse slope (lowest in the centre) also provides better air movement. This style of roof must slope to one end to discharge rainwater. Birds must be protected from the radiant heat from a metal roof by a ceiling of anything that is a barrier to heat rays (anything that would provide shade from the sun such as cardboard, cloth, dried leaves, etc.). Trees close to buildings may be a disadvantage for large pens because they interfere with air movement.

- *Floor:* A wire or wooden slat floor for broilers, or cages raised 1 to 3 m above the ground improves air movement.
- Thick clay walls of traditional buildings usually remain cool on the inside and provide conduction and convection cooling for the chickens.
- Fans can be used on commercial farms to increase air movement. Tunnel ventilation is most effective.

Evaporative Cooling: Evaporating water removes heat from the air. This technique can be used to help keep chickens cool:

- Sprays or misters can be used inside a pen. These are best with open floor (no litter) pens.
- Water can be dripped over the roof or from the sides of the pen. Water soaked pads or screens will increase evaporative cooling. Air should move through the pads to the chickens. Fans can be used to draw air through wet screens or other material.

Feeding and management:

- Feed consumption increases heat production by the body. To prevent heat stroke, feed may be withdrawn 2 to 3 hours after daylight and provided again in the evening.
- Increased water consumption helps cool the birds. Cool water is best. Water must be freely available in the parts of the pen where chicks gather to try to keep cool. These areas are usually towards the direction of the air movement (wind or fan), or over open floor areas, or away from places that are hotter because of radiant, conductive or convection heat.

Genetics: Some strains and breeds of chicken are less affected by heat. Fat chickens are most at risk. Chickens with fewer feathers (naked neck) and large combs are less at risk.

Waterfowl

Waterfowl, particularly ducks, are very popular in many Asian countries. They are good foragers and can scavenge better than chickens. Drinking water and shade must always be available. Ducks prefer to forage around ponds and rivers and can get much of their nutrition from water, animals, plants and insects.

Ducks: Ducks can be kept for meat or egg production. White Pekin, Rowen and Muscovy are best for meat. Kharki Campbell and Indian Runner are egg laying breeds. Muscovy ducks are different from other breeds. They can fly a short distance; egg incubation is longer (35 days instead of 28) than other breeds; the male is larger than the female and has unfeathered wattles and carucles on the head; there are white as well as coloured varieties. Muscovy ducks are also useful for fly and other insect control. Some local breeds of ducks are also kept and wild ducks may be domesticated.

Ducks may be allowed to scavenge for their food or may be confined. Flocks of ducks that are confined would be fed similarly to confined chickens. Ducks prefer to have a pond in the yard. The pond could also be used for fish. Free ranging ducks will grow and produce better if they have access to a pond, lake or river.

Free ranging ducks can be herded the way sheep and cattle are herded. The herding can be done by a person or person with a dog. Duck flocks can be herded slowly for several kilometres to scavenge on fields that have been harvested or in orchards and some fields of vegetables. Survival rates are improved by providing feed and protection to young ducks for 2 to 3 weeks.

Males are hard to distinguish from females in some breeds. One male (drake) is usually mated with 5 to 8 females if breeding ducks are kept in confinement. Ducks may be allowed to incubate their own eggs, eggs may be collected and hatched under a hen, or eggs may be incubated artificially.

Nest boxes for egg laying strains or breeders should be in a protected area and raised 10 to 15 cm from the ground or floor.

Geese: Geese are kept for meat and feathers. In some places geese are also used as alarm systems to make a noise when strangers or predators approach.

There are several breeds of domestic geese and some wild geese can be domesticated easily and used as food. Geese are excellent foragers and survive well on young legume and grass pastures and on many other young weeds and plants.

On a pasture the plants should be kept clipped or pastured every two weeks by cattle for several days to remove the older plants. Geese can be used to remove weeds from orchards, gardens and field crops such as cotton, sugar beets and strawberries. They will also control grass and weeds on rivers, marsh and swamp land.

Geese can be aggressive and can defend themselves and their goslings against small predators.

6

Poultry Feeds and Feeding

Definitions and Terms Used

Nutrition: The process of digesting, absorbing and converting food into tissue and energy. Also, the study of this process.

Nutrient: A substance that can be used as food. Some people use "food" for nutrients eaten by people and "feed" or "feedstuff" for animals.

Digestion: The process of changing food to a form that can be absorbed from the digestive tract by the body tissues (mainly the intestines). In the digestive tract this is done by enzymes and other material produced by the digestive tract which break down the food into small, simple (molecular) components.

Digestion by microorganisms: Some bacteria and protozoa produce enzymes and other material that break down cellulose, and fibre, etc. that non-herbivores (non-grass eaters) like chickens cannot digest. There is some digestion by microorganisms in the cecum of chickens. Enzymes produced by bacteria can be added to some feeds to improve digestion and absorption.

Metabolism: A chemical reaction that takes place in the tissues and organs of the body in which the food that has been digested and absorbed is changed either into energy or building blocks for the body. Energy is the power produced by the food. It is the fuel, like petrol for cars, on which the body runs and, like burning wood or petrol, metabolism requires oxygen and produces heat and waste material (carbon dioxide and water).

Basal metabolism, basal metabolic rate (BMR): The amount of energy (fuel) required to keep the body alive and operating without activity, growth or production.

Energy: Energy is the amount of power produced when food is metabolised. Energy is measured in heat units (calories or joules). In nutrition the kilocalorie (kcal) equals 1000 gram calories. A gram calorie is the heat required to raise 1 gram of water 1BC (14.5 to 15.5BC) (1 kcal equals 4.184 kjoules).

Metabolisable energy (ME): The amount of energy available to be used for maintenance, for production of body tissue (for growth and replacement), activity and egg production, when a food material or feed is eaten. It includes the heat lost during metabolism. The ME of a feed ingredient (individual feed) may be used to indicate the nutritional value of that ingredient. Feed ingredients or feeds are rated as high or low energy. ME depends on the quality of the feed and on the % dry matter. Good maize (corn) at 85% dry matter (15% water) has an ME of 3300 kcal/kg. The ME for barley is 2700 kcal/kg. Fats may have an ME of 9000 kcal/kg. If chickens are fed a low energy feed they will eat more feed, if it is available, to get the required energy. In monogastric animals, like chickens, energy comes mainly from carbohydrates and fats since fibre containing cellulose cannot be digested.

Organic: Organic compounds are defined in nutrition as animal or plant material containing carbon.

Carbohydrates: Carbohydrates are organic compounds and a source of energy for poultry. Simple carbohydrates are made up from sugars: glucose from maize is a monosaccharide. Lactose from milk, sucrose from sugarcane or sugar beets and in smaller amounts in many plants, particularly in their fruits, seeds or roots are disaccharides. Complex carbohydrates (polysaccharides) are made up from combinations of sugars. Some, like starch (from cassava, maize, wheat, etc.) can be digested by chickens while others (cellulose, a structural carbohydrate) cannot because animals do not have the enzymes to break down (hydrolyse) cellulose so that it can be absorbed. This digestion is done by bacteria and protozoa in herbavores.

Fats (oils, lipids): Fats are a source of energy and in some cases fat soluble vitamins. Like carbohydrates, they are organic compounds made up from carbon, hydrogen and oxygen which form fatty acids. Because they are higher in hydrogen and lower in oxygen than carbohydrates, fats have a higher energy value (ME) than carbohydrates. Fats that are high in unsaturated fatty acid are liquid at room temperature and are called oils. Vegetable oils (canola oil, corn oil, cottonseed oil, soybean oil, etc.). Most essential fatty acids (required for health & growth) can be produced by the chicken from other food; but linoleic acid must be present in the diet of chickens for proper growth and egg production.

Proteins: Proteins are organic compounds used to build the cells, tissues and organs for the body. They are made up from many amino acids some of which are essential for growth and production. Proteins contain about 16% nitrogen so the amount of protein (crude protein, not the digestible or usable protein) in a feed can be estimated by measuring the nitrogen content. Excess protein (above what is required for growth or production) is used as energy and the nitrogen excreted in the urine. During digestion protein is broken down into individual amino acids for absorption. In the body these are reassembled to make body tissue or egg protein. Birds can make some amino acids from other protein but many amino acids are essential, that is they must be present in the diet. Proteins from animal sources (meat, milk, insects, larvae, etc.) contain the essential amino acids. Most plants are low in protein and vary widely in essential amino acid content. Most are low in one or more essential amino acids. Some legume plant seeds such as beans are high in protein.

Vitamins: This term describes a variety of essential nutrients that are not similar to one another, except that they are essential in the diet, but only required in very small amounts. They are used as metabolic regulators. They are either water soluble or fat soluble. Vitamins or their precursors are present in small, but variable amounts in some feeds. Most vitamins are manufactured (synthetic) for use in commercial feed. Precursors of vitamin D3 can be produced in the skin by sunlight. Vitamin C is produced by bacteria in the intestine in chickens.

Minerals. Minerals are Not Organic: They are chemical elements. Those required in small but significant amounts (calcium, phosphorus, sodium, potassium, magnesium and chlorine) are major minerals. Essential trace minerals are iodine, cobalt (as cobolamin), iron, copper, zinc, selenium and molybdenum. Most minerals must be added to the diet for good growth and egg production. Laying hens need a ration with 3% calcium to make egg shells. This can be supplied free choice as small pieces of bone-meal, coral, sea-shell or limestone. Phosphorus in plants is present as phytate phosphorus and is only partially available unless phytase enzyme is added to the feed.

Water: Water is an essential nutrient. Water is the major part of animal tissues and all body functions require water.

A source of clean, cool water should always be available free choice.

Antibiotics: Antibiotics are drugs made from bacteria or fungi that are used as medicine to prevent or treat bacterial disease. They are sometimes used as growth stimulants in chickens.

Probiotics: Probiotics are bacteria or products produced by bacteria that encourage the growth of "good" bacteria (those that prevent the attachment and or growth of disease causing bacteria) in the intestine.

Chemotherapeutics are chemical compounds used as medicines.

Vaccines are preparations of live organisms, (mainly viruses) used to produce immunity (defence) against disease causing viruses. They stimulate antibody production by the chicken to protect against the virus material in the vaccine. Because they are live and must multiply in the body to be effective, they can be used as a spray, in water, eyedrop, wing web injection or subcutaneous injection.

Bacterins are preparations of killed organisms (bacteria, mycoplasma, virus, etc.) for subcutaneous or intramuscular injection. They also stimulate antibody production.

Antibodies are very small particles produced by lymphocytes (type of blood cell) that circulate in the blood to help the body defence systems stop infection by specific virus or bacteria that had stimulated the lymphocytes to produce the antibody. They can also be passed in the egg from the hen to the chick (maternal antibody) and protect the chick for 1-3 weeks.

Cereal Grains: Cereals are the grains grown specifically from human and animal food. They include wheat, millet, rice, maize (corn), sorghum (milo, kafir or guinea corn), barley, oats, etc. Grains provide the main source of energy in commercial poultry feed. Protein in cereals are low (8 to 12%) and the quality (level of essential amino acids) is poor.

By-Products: By-products are the parts of a grain, oilseed, or animal that is being prepared for human food, that is not used for human food. Examples are wheat or rice bran or animal viscerae that may be fed to chickens.

Complete Feed: A prepared feed that contains all of the nutrients for the best growth or egg production for the flock for which it is being used is called a complete feed. These feeds are usually prepared commercially by a feed manufacturer. Broiler chickens receive high protein (21 to 23%) starter for the first 14 to 21 days, grower to day 28 or longer and finisher, that is lower in protein and higher in energy, until ready for market.

Supplement or concentrate is a prepared feed that is intended to be added to or mixed with other feed material to improve the nutrient balance of the final feed.

Premix: A premix may contain a variety of specific ingredients such as vitamins, trace minerals, amino acids, or medicine that is to be mixed with other feed to supply essential or important elements that may be missing. Essential nutrients are sometimes added to the water if they are water-soluble.

Antinutrients: Some feed ingredients and potential feeds contain factors that inhibit the digestive process causing reduced growth, diarrhea or pasting. They limit the amount of some feed ingredients that can be added to the final feed. The antinutrition factors in some feed material such as beans can be destroyed by heat (cooking).

Phytotoxins are toxic or poison substances found in plants used as feed for chickens. The toxic material can be in the seed (castor bean) leaf or stem or in the root or tuber (cassava). At low levels some might only reduce growth rate or have no effect. At higher doses they might cause illness or death. Toxic weed seeds may contaminate grain that is being harvested making that feed toxic.

Mycotoxins are poisons produced by moulds growing in food material. Various fungi produce different toxins. Aspergillus fungi produce aflatoxin, one of the most serious in hot climates. The fungi can grow in crops in the field, in seeds in storage or in prepared feed for chickens.

Nutritional Requirements of Poultry

Poultry require carbohydrates, fats, protein, vitamins, minerals, and water. Most nutrients provide both energy (as carbohydrate or fat) and protein. Feedstuffs are usually classed as being primarily a source of energy or protein. Fats and oils do not provide protein. Energy and protein are more efficient if they are available in the proper ratio. Excess protein is used as energy with increased waste excretion. Growing chickens should have 16-24% protein; growing turkeys, 24-28% protein; for egg production 15-17% protein; for maintenance (no growth or production) 10-12% protein. The highest requirement is in the first 2 to 3 weeks and is higher in young leghorns than broilers.

The protein level can be lower if the essential amino acids are all present at the proper level. The vitamin and mineral levels are higher for growing birds and for egg production. More rapid growth requires more added vitamins and minerals than slow growth. In slow growing birds more of some of these essential vitamins and minerals are available from the feed ingredients because the amount of feed required for basal metabolism is greater.

Some feed and water is required just for survival. The BMR varies directly with the weight of the bird. If the feed energy required to supply the basic energy need is not available, the bird will lose weight, get thin and die. Some protein, fat, vitamins and minerals are also required for basic survival. Nutrients available above the basic level can be used for growth in young birds and egg production in adults. Excess energy is converted to fat and stored in the body. It may be used when energy from feed is lacking as in the hump in camels and tail fat in fat-tailed sheep.

The rate of growth in young birds or egg production in adults is controlled by:

a)   the genetic potential (commercial broilers compared to layers)

b)   the amount of energy and protein (and protein quality) available above the level for the BMR. The feed required to maintain BMR does not add to growth or production. The maintenance requirement for a slow-growing broiler is the same as a fast-growing broiler of the same weight. The feed maintenance requirement for a hen laying 2 eggs a week is the same as a hen of the same weight laying 6 eggs a week. Sufficient vitamins and minerals must also be available.

If the temperature is below the birds comfort level, additional energy is required to provide internal heat by increased metabolism to maintain body temperature. Activity also increases metabolism and birds that are allowed to run outside or that scavenge for their feed have a higher requirement for energy.

Chickens that get all their nutrients from scavenging may eat an excess of protein, if insects, worms, larvae, etc. are available. They might benefit most from supplemental feeding of energy in the form of carbohydrate (cereal grains, etc.). Fenced or backyard poultry fed household or garden waste may lack both energy and protein for good growth or egg production.

Laying hens should have calcium available free-choice, even if calcium is being added to feed. Chickens that eat whole seeds, grains, vegetable material or fibre must have insoluble grit (granite grit) or small stones in their gizzard to grind the hard or fibre material. Birds that forage pick up their own grit. Poultry kept indoors must have grit supplied. Small chicks require small stones 2 to 4 mm. Hens .5 to 1.5 cm. Limestone particles are not satisfactory as grit.

The amino acids that growing poultry require are: arganine, cystine, lysine, methionine, tryptophan, histidine, isoleucine, leucine, phenylalanine, threonine, tyrosine, valine.

The following vitamins should be added to a prepared feed: vitamins A, D3, B12, riboflavin, pantothenic acid, niacin, folicin (folic acid), thiamine and biotin.

Commercial Nutrient Sources

In countries where complete feeds are prepared by feed manufacturers, cereal grains and oilseed meal are the major ingredients in the feed. Depending on cost, availability and age or type of poultry or water fowl, some of the following products are also frequently used: by-products of cereal grains processed for human use, animal processing waste, fishmeal, waste oil from restaurants (restaurant grease), yeast, alfalfa meal, distillery or milk factory products, etc. Vitamins and mineral are also added in the correct proportions. The ingredients are ground (if required), mixed and may be pelleted.

Ingredients most frequently used:
- for energy: maize (corn) up to 65%
- sorghum (milo) up to 45%
- wheat up to 25%; with enzyme up to 45%
- wheat by-products (bran, shorts, screenings) up to 15%
- rice up to 15%, rice by-products (bran, polishings) up to 15%
- barley up to 15%; with enzyme up to 35%

- molasses, up to 5% after 2 weeks
- for protein: soybean meal up to 30%
- soybeans up to 15% (heated to remove antinutrients)
- canola meal or whole seed up to 10%
- corn gluten up to 15%
- peas, lupin, flax up to 10%; flax (linseed) meal - 15 % (20% in layers)
- safflower meal, sunflower meal up to 10%
- meat meal, fish meal up to 10%
- blood meal, feather meal up to 2%
- Fats and Oils: tallow, lard, coconut oil, palm oil, after 3 weeks up to 5%
- poultry fat, fish oil, restaurant grease up to 3%, after 3 weeks 6%

Minerals: Calcium can be added as limestone, as prepared products such as dicalcium phosphate or as oyster or other marine shell. Growing birds require about 0.9% calcium in the diet. Hens in lay require 3 to 4% calcium in the diet so that calcium is available for egg shell production.

Phosphorus can be added from manufactured products such as dicalcium or monocalcium phosphate. Phytate phosphorus (an organic phosphorus) is present in small but varying amounts in plants feed ingredients. This phytate phosphorus is only 25-50% available unless phytase enzyme is added to the ration. Growing birds require about 0.4% phosphorus, adults about 0.3%. Calcium and phosphorus levels must be in balance since high levels of one cause deficiency of the other.

Sodium and chlorine are usually added as salt (NaCl) at about 0.25% of the ration (2.5 kg NaCl/1000 kg).

Magnesium and potassium are usually present in adequate amounts in the feed ingredients. Some ingredients such as soymeal and molasses are high in potassium. Minerals may be present in high or moderate levels in animal by-products. If animal by-products are being fed the minerals present in this feed should be used in the calculation of the amount to be added. High levels of sodium are toxic to birds. Saline water is also toxic for birds.

Trace minerals and vitamins are required in small quantities. They are not usually added as separate ingredients. They are usually bought as a prepared package or a premix. A specified amount of this premix can then be added to each tonne (1000 kg) of feed. To make sure of an even distribution in the final feed the vitamin-mineral premix may be mixed with 25-50 kg of cornmeal or other large particle ingredient that will help ensure an even mix. This is then added to the larger mixer after the cereal and protein portions have been put in. The calcium, phosphorus and salt should never be added to the vitamin premix. The mixer should be grounded.

Alternative Feeds

There are many vegetable and animal products that can be used as feed ingredients for poultry. Poultry cannot digest fibre found in roughage, seeds, grains and fruits. Products that are high in fibre can make up only a small percent of the feed. Since many plant products also contain antinutrients the amount of these feed stuffs that can be fed is also limited. Products that contain a high percent of water (kitchen and garden waste, weeds, tree and plant leaves, fruit, worms, larvae, fresh fish and animal waste, water animals and insects, milk products, etc.) can be given free-choice. What is not eaten must be removed so the chickens will not eat mouldy or spoiled material. Most feeds can be sun-dried to reduce the water content. They can then be ground into meal to be mixed with other feed, or cut up into chips that the birds can eat.

Many fresh (undried) foods such as garden and kitchen waste, fruit, plants and vines can be fed free choice. The best way to preserve foodstuffs is by sun-drying. It is important to avoid spoilage. Some types of foods can be preserved by anaerobic (no air) fermentation (making silage) where the fresh food ferments and becomes acid. The acidity prevents the growth of spoilage organisms. With plant material this may be done in a large plastic bag kept in the shade. Milk may be preserved for short periods by allowing it to sour (go acid) or by adding acid material.

Seeds and grains that contain antinutrients and toxins must be used with care since levels of antinutrients and toxins vary. Several varieties of raw or unprocessed peas and bean should not be used together (don't use 5% of each) because they contain similar antinutrients and toxins.

Proteins are fed for the nitrogen and amino acid content. Some amino acids are essential and must be present in the feed or added as synthetic amino acids. Poultry cannot make them from other amino acids or nitrogen. Feeding a variety of protein sources may supply more of the essential amino acids.

Poultry require both sodium and chlorine. There is some sodium and chlorine in feedstuffs, but both must be added to the feed, usually as salt. The recommended level is 0.25 to 0.3% added salt (2.5 to 3 kg per tonne). If feed ingredients or drinking water contains salt, the amount of added salt must be reduced. Poultry can be poisoned by the sodium in salt and young chickens are very susceptible to high sodium. Chicks are frequently poisoned by salt in the drinking water. Water for young chicks should not contain salt and many chicks die from salt poisoning at levels below 0.1% (0.9% is the level found in animal tissue). Chicks can be poisoned by eating salted fish and particularly by the brine used to preserve fish.

Young chicks can also be poisoned by high calcium levels in the feed. Feed for layers contains 3 to 4% calcium. That level can cause kidney disease in young chicks.

The Importance of Water

Water is an essential nutrient. All body functions and processes require water. Water is the material in which all other nutrients are carried in solution. The tissues and cells in the body are made up mainly of water. Water is also present between cells so that material can be moved between cells. Dry feed contains only 10-15% water and poultry need about twice as much water as feed (2 gms of water for each gm of dried feed).

Water should be clean and free from chemicals and minerals. It should not contain harmful parasites or bacteria. Water should be easily available free-choice. If water consumption is restricted, growing chickens will grow more slowly. Broilers that get only 2/3 or 1/2 of the amount of water they want will eat less feed and grow at only 3/4 of the expected rate. In cool climates water restriction is used to slow growth in breeders or to reduce the ascites syndrome. If water is not freely and easily available adults will lay fewer eggs and may suffer from kidney disease (as do people that do not consume sufficient water). Layers that go without water for a day or two will stop laying and may take 2-3 weeks to recover. Water in a trough or cup must be deep enough to allow the bird to drink properly. A depth of 2 cm is adequate, unless poor beak trimming has resulted in long lower beaks. The drinkers must be easy to reach and housed poultry must not have to walk too far to get water; not more than 2 metres for broilers and 4 meters for adults.

In hot climates water availability is even more important. The amount of water required becomes much higher as the temperature rises above 25BC. Birds cool themselves by water evaporation from their respiratory system and lose water when they mouth breath. In hot climates cool water is better than warm water. For large groups of poultry the water pipes to the poultry sheds should be 40 to 60 cm underground to cool the water as it flows to the chickens. Above ground pipes and emergency tanks should be shaded from the sun. To keep the water in emergency tanks fresh it can be used for 3 to 5 hours in the early morning. Chickens will move to cooler parts of the pen when the temperature is high, particularly if there is a breeze or fans. It is important to have water easily available in these areas because the birds may not go back to hot areas to drink.

Nipple drinkers help conserve water and avoid spillage, but chickens drink less water from nipples. Nipple drinkers require careful management. The pipes must be level, at the correct height (at eye level for chicks, just above eye level for older birds) and the water pressure must not be too high for the birds to operate the nipple. The water column that controls pressure should be about the same height as the chickens using the system (6 cm for a 6 cm chick).

Newly hatched chicks should have water before or at the same time as they receive their first feed. It must not be in an open pan or trough that the chicks can get into. Wet chicks lose body heat quickly and may die. Water jugs inverted in pans with a narrow water source are satisfactory as are raised narrow troughs, bell-type drinkers and nipple drinkers. Water jugs, troughs, etc. must be cleaned at least once a day.

Water Problems

Water containing physical particles or dissolved mineral material may interfere with automatic drinkers. Particulate material and iron may be removed by a sand filter and settling tank but dissolved minerals and algae growth in the system are more difficult to control. Regular cleaning and flushing may be necessary. Copper sulphate solution may control fungi and algae growth, but it can be toxic for poultry. Poultry may be poisoned by minerals or chemicals in the water. The sodium in saline (NaCl) water is the most frequent problem. Young birds are very susceptible to sodium toxicity. Sodium above 500 ppm (0.05%) in the drinking water may cause death in young chickens and turkeys depending on the sodium level in the feed. The young birds usually die from oedema and ascites syndrome. Wet droppings or diarrhea would also occur. Salt in the feed can be reduced to avoid disease from salt in the water up to about 1000 ppm (0.1%) of sodium in the water. Saline water above 1000 ppm should not be used for broilers up to 21 days of age even if no salt is added to the feed. High levels of sodium will kill adult chickens by causing diarrhea and dehydration.

Sodium may be present in water as sodium sulphate ($NaSO_4$) or sodium bicarbonate ($NaHCO_3$). Sulphate may be present as magnesium sulphate ($MgSO_4$) and cause diarrhea and death. All sources of sodium in the feed, whether added as salt or present in the feed ingredients, (particularly animal protein) and water are additive in causing sodium toxicity. If sodium is too high in the feed it will cause disease as it does in the water. Sodium, however, is an essential nutrient and poultry require some sodium to grow and produce eggs.

Nitrate in feed and water may be toxic at high levels but can reduce growth at 50 ppm in water. A variety of other naturally occurring minerals and chemicals may be present in water and may cause problems. Surface water may be contaminated by farm pesticides, fertilizer or by industrial chemicals.

Brooding and Growing Chicks and Poults

Hatched chicks should be lively, firm and healthy. They should be uniform in size and have a well-healed navel. They should be started as soon as possible after removal from the incubators. The sooner they are provided a warm area, feed and water, the better the rate of success. Chicks must not be chilled or overheated at any time. If chicks do not receive feed and water within 18 hours after hatching they should get water for 1-2 hours before they are given feed.

Systems of Brooding

All pens and equipment should be ready, tested and operated before the chicks arrive.

Natural Brooding: For small numbers, a broody hen can handle 14-15 chicks. Protection from predators and rain is important. Chicks should get some commercial ration or other feed for the first 3-4 weeks. Sour milk is good for young chicks. A little care over the first 3-4 weeks will often double or triple the survival rate.

Warm Room Brooding: In this system, the whole pen is held at a temperature of 30-32BC both day and night. The temperature is lowered about 1.5BC per week until the ambient temperature is reached, but should not go below 21BC until 6 to 8 weeks and not below 18BC until 10 to 15 weeks. Light should be over feed and water.

Circle Brooding: In this system, brooding areas of the pen under the heat reflector are heated to around 35BC by the use of reflectors over some form of artificial heat, such as gas heaters. Chicks are enclosed by cardboard chick guards, (not more than 400 chicks per circle) that allow them to move away from the heat source, but keep them close to feed and water. Light should attract chicks to the heat source. Temperature in the rest of the pen can be allowed to drop to 20 or 15BC.

Heat may be supplied by hot water or hot air with the heat source usually oil, gas, coal, wood or electricity. For small flocks of up to 75 chicks, a heat lamp may be all that is required. Temperature readings should be taken 5 cm above the litter (or at chick level). Make sure the thermometer is shaded from radiant heat.

With circle brooding, the feed and water should be placed at the edge of the heat reflectors, which is 60 to 70 cm above the floor. Comfortable chicks will be at the feeders and waterers and spread uniformly over the pen area. When chicks feel cold, they will crowd under the heat source or pile up.

If the pen is too warm, the chicks move away from the heat with wings spread and/or are panting.

Uncomfortable chicks will be noisy:

(a) Floor space - Allow 1m2 per 75-150 chicks under the heat source and at least 1m2 per 25-50 chicks within the chick guard. Many producers use a chick guard of corrugated cardboard 30 cm in height placed 2-3 m back from the heat source for the first 5 to 7 days. The guard should be removed when the birds can fly over it. Broilers at 6 weeks require 1m2 per 9 to 12 birds.

Leghorn chicks for egg production are often started in large (1 x 2m) or smaller growing cages with warm room brooding, or in hot climates with a heat source over each cage. Small growing cages may be about 50 x 50 cm with 24 chicks started in each. After 4 to 5 weeks, half the chicks are placed in another cage of equivalent size, to double the space per bird and split again to 6 per cage at 8 to 10 weeks (400 cm2 per bird).

(b) Feeders - Chicks should be given 5 (Leghorns) -8 (broilers) cm of feeding space. Feed is usually put in shallow boxes or egg flats on the floor or cage bottom for 4 to 6 days as well.

(c) Waterers - Good water is important. In large flocks, some form of automatic water system is usually installed. In a float controlled trough, 2 cm per bird is considered sufficient. One hanging fountain per 80 to 100 birds, one smaller cup per 50 birds, one nipple for 10-12 birds, will generally be adequate.

For small groups, three 4L or larger drinkers should be provided per 100 chicks, and more added as required. These should be cleaned and refilled daily. Feed and water should be within 1.5m of all chicks. Water should be clean and free from toxins and chemicals. Salt as $NaCl$ or $NaSO_4$ is particularly dangerous in young chicks up to 21 days. Total sodium ($Na+$) in water should not be above 300 ppm for broilers and 600 ppm for Leghorns or local strains.

Feeding

For best results, a commercial ration is recommended. If home grain or other feed is available, it may be fed starting at day 8 in broilers and day 15 in Leghorns. Replace 5% of commercial ration with whole grain or other feeds and increase by 5% a week in broilers (5% every 14 days in Leghorns) to a maximum of 50%, or a commercial concentrate may be diluted with whole or ground grain or alternate feed stuffs. When grain is fed, supply insoluble grit sprinkled on the feed at least once per week. Don't overfeed grit.

Chickens raised for meat should be given all the feed they will eat (unless they are being raised at high altitude). Layers may have to be restricted to prevent them becoming too fat. Breeding flock must be monitored regularly and weighed weekly to make sure they are uniform and not overweight. Severe restriction may be required in broiler breeders.

It is advisable to include an anticoccidial drug in the feed of birds grown on the floor. This should be fed continuously throughout the growing period up to market in broilers and to 12 weeks in breeders or Leghorns. Coccidia vaccines are also available.

Cannibalism (Feather Pulling - Vent Picking, Toe Picking, Etc.)

Birds are cannibalistic by nature and when confined may start picking.

Chicks should be beak-trimmed at the hatchery, or at the farm using an electric trimmer. Broilers should not need trimming but watch for signs of picking and beak-trim if necessary.

Other steps to consider are:
- decreasing the light intensity;
- decreasing pen temperature;
- provide more space;
- use red light bulbs or place red plastic over windows;
- feed whole oats if the flock is over 4 weeks of age;
- add 1 g salt per L of drinking water if flock is over 6 weeks of age.

Once started, cannibalism is difficult to control.

Trouble Shooting

- Chicks should be well spread out and at the drinkers and feeders
- If noisy and crowding the walls or chick guard they may be too warm
- If noisy and crowding the heat source they may be too cool or drafty
- If crowded around feeders or drinkers, check supply and space per bird.
- If chicks are huddling or piling in small lots they may be too cold, too hot or sick.
- Litter condition
  - too dry, may cause respiratory problems - spray litter to increase humidity.
  - too wet, ammonia and coccidiosis or leg problems occur - check ventilation and bird floor space.

- Insufficient feeder or waterer space, disease, poor lighting, etc., will all affect uniformity.

Poultry Health Management for Commercial Poultry

Disease Prevention

Diseases caused by infection with a living microorganism such as bacteria, virus, mycoplasma, parasite, etc. are infectious diseases. Most infectious diseases are also contagious, that is, they spread from one chicken to another but a few, like Staphylococcus infection, and aspergillosis, are not.

Prevention by Sanitation

Sanitation is used to reduce the numbers of disease organisms which the chicken contacts to the level where they will no longer cause disease and to provide a clean, healthy environment. This can be done by cleaning and disinfecting, by adequate ventilation to reduce the number of organisms in the air and by reducing contact with other chickens by keeping them in cages.

Sanitation affects all levels of the birds' environment:

- *Building and Equipment:* Cages are intended to keep birds out of contact with their droppings, spilled feed and water, etc. that are a source of bacteria and bacterial growth. Fly control is important if this method is to be effective. On floor operations, an adequate amount of dry litter helps control bacterial growth.

  At cleanout, blow or spray dust from walls, ceiling, rafters, fans, etc. Wet down and remove litter, clean the building and equipment using detergent and water. Follow with a good disinfectant. A dirt flood cannot be disinfected. Litter should be piled or spread at least 100m from the poultry buildings (1 km is prefered).

- *Feed:* Between flocks, clean out feed in the troughs and bins both inside and outside the building and remove it from the building. It can be held to feed to chickens of a similar age or sent for reprocessing. Outside storage bins must be checked to make sure spoiled or moldy feed is not adhering to the inside and that the bin is empty. Feeders should be constructed and set so that feed is not spilled or billed out into the litter.

- *Water:* Open troughs are a source of contamination from the nasal and oral secretion and feces, etc. and must be cleaned regularly. Nipple drinkers are much cleaner but pressure must not be too high. Tremendous numbers of bacteria can build up in ponds and some wells, in water lines, and in drinking cups or troughs, so a periodic or constant water-sanitising programme and descaling may be necessary.

- *Air:* Clean, germ-free air is a very important part of a healthy environment. A good ventilation system to dilute and carry away dust and microorganisms and keep the pen dry is part of a sanitation programme.

- *The Caretaker:* Workers can carry infection to birds on their hands, clothes, boots, and equipment. They can spread germs from one group to another on the farm or bring germs in from outside the farm. If the same person is looking after more than one building, outer clothing and footwear should be changed between buildings and hands washed in disinfectant. If possible, separate workers should care for chicks, young birds and adults.

Good isolation requires shower-in and no contact with other chickens or other people who have chickens or work with chickens. Sanitation is a method of eliminating or reducing the number of disease causing organisms from contacting the birds. Although one germ may not cause disease and healthy birds may resist hundreds of germs, sanitation cannot be relied on to control serious contagious diseases. It must be used in conjunction with isolation and sometimes vaccination for problems such as Newcastle disease, avian influenza, infectious laryngotracheitis, fowl pox, etc. However, sanitation is important in maintaining the general health of a flock and is an important part of good flock management.

Prevention by Isolation

This method of disease control is simple. Stop the microorganisms that cause disease from contacting the chickens. When the microorganisms which cause a disease are eliminated from an area or country, the disease is said to be eradicated.

Whether or not a disease can be prevented by isolation depends on:

(a) Where the microorganisms that cause the disease live. Some organisms (like the ones that cause necrotic enteritis, staphylococcosis, coccidiosis, *E. coli*, etc.) are widespread in the environment and it is difficult to prevent birds from contacting them. Some organisms are found primarily or only in chickens (sick birds or healthy carriers) and live outside the bird's body for only a short time (mycoplasma). Others may live for days, weeks, or months in the environment, depending on the organism and the conditions (moisture, temperature, etc.).

- The way the disease organism is spread. Most disease germs are spread by direct contact, that is bird to bird. If the disease germs get on poultry fluff or small, air-borne particles, they can be carried by the wind, but this usually only over short distances. Organisms are frequently spread by mechanical transfer by carrier objects such as people, eggs, egg cases, trucks, feed, water, rats, dogs, insects, etc. Two or more separate objects may be involved in the spread of organisms from one flock to another (chickens to eggs, to egg cases, to people, back to chickens). Some animals, and certainly wild birds, can become infected with a disease and spread the germs. Insects are a necessary part of the spread of some diseases (leukocytozoon, plasmodium, etc.). Some disease organisms are spread through the egg from the adult to the chick (vertical spread). Some disease germs are spread venereally or by artificial insemination in semen (mycoplasma).

Diseases that are easier to control by isolation are those where the causative agent survives outside the body for only a short time. If the agent is egg-transmitted, the breeder flock must also be free of the disease. With some diseases a regular monitoring programme may be required to insure that the flock remains negative.

Isolation is not only the cheapest but the best way to control many contagious diseases and good isolation is equivalent to a quarantine.

Important Features of an Isolation Programme

Workers must be trained to carry out the sanitation and isolation procedures. They must also understand why the rules are set up and why disease prevention is important.

- Have only one age group on the farm (an all in, all out programme). Buildings over 100 meters apart can be treated as separate units if proper isolation and sanitation procedures are followed.

- Obtain chicks or replacements from a disease-free, adequately isolated, single source or raise replacements in a different area with separate caretakers.
- Have no neighbouring poultry buildings or free ranging chickens within 300 meters.
- Clean and sanitise buildings and equipment between crops. (wet down litter before removal to protect neighbouring poultry and do not store or spread litter near poultry buildings).
- Screen buildings against wild birds and keep out rats, cats, and dogs, and control insects. Remove dead birds from the pens at least twice a day and dispose of sick and dead birds at least 100m from the poultry buildings. Make sure dogs, cats and wild birds or animals cannot drag or carry dead chickens onto the farm.
- Limit the movement of workers from one building to another.
- Bring in only new or sterilised egg cases and flats.
- Make sure employees do not keep poultry or pet birds or come in contact with free-range chickens or their droppings and do not visit other poultry farms.
- Keep out visitors (particularly those who may visit other poultry farms) and provide boots and protective clothing for persons entering the poultry area.
- Disinfect necessary vehicles (feed trucks, etc.) and restrict them to the loading and unloading areas which should not be near the building entrance. Keep the driver in the truck or provide boots and coveralls.
- Make sure poultry service crews disinfect equipment, shower, and change clothing before entering the poultry area (except at cleanout).
- Shower and change clothing after taking chickens to market or meeting with poultry workers from other farms.

Prevention by Vaccination

Poultry have a good immune response to many diseases and to vaccination. They also pass immunity to offspring through the egg. Breeders require a special vaccination programme.

Marek's Disease (MD): There are several types of vaccine. All are live and given by injection. If vaccine is available, all chicks should receive MD vaccine immediately after hatching. Frequently two types of vaccine are given together. A second vaccination at a later date is not required. Chicks must be kept in a clean pen, away from other chickens for three weeks.

Newcastle Disease (ND): There is only one serotype of NDV so proper vaccination protects against all pathotypes of virus. Maternal antibody is provided for several days by immune hens. In countries where velogenic virus is endemic, chicks should be vaccinated before day 7 at day 21, 35 and 48 using more immunogenic (or virulent) vaccine (LaSota) for the 3rd and 4th vaccinations. Vaccinations should be repeated at 45 to 60 day intervals in places where ND is endemic. Properly administered live spray vaccine is best in adults. Eye drop administration is best for chicks. Either eye drop or spray can be used after 8 weeks. Vaccine in feed is available for village flocks. Killed vaccine may be given by injection.

Infectious Bronchitis (IB): There are many serotypes of IBV. Although there is some cross protection, new serotypes appear regularly that can evade the immunity of some vaccines. Commercial layers should be vaccinated as recommended locally. Broilers may also require protection. Live vaccines given in water or by spray are best in growing birds. Killed vaccines usually in oil may be used for adults. Some serotypes of IBV cause kidney disease.

Infectious Bursal Disease (IBD): There is one major serotype of IBDV but there are many pathotypes. As with ND, mild strains of vaccines will not protect against highly pathogenic strains of virus. Maternal antibody is provided for 7 to 14 days by immune hens. Where pathogenic strains of virus are present, moderate strength vaccine should be given at days 8, 14, and 28. If maternal antibody is not uniform, mild vaccine can be given at day 1. For maternal immunity killed vaccine in oil (often with ND & IB) may be given twice at 16 & 18 weeks and may be repeated at 35 to 45 weeks if necessary.

Avian Encephalomyelitis (AE): Epidemic tremors. Only breeders must be vaccinated but egg layers should be done as well to prevent a drop in egg production. Vaccination is done at about 12 weeks by water or wing web stab (live vaccine).

Viral Arthritis (VA): The arthritic form of reovirus infection is rare in many countries. Where the disease occurs broiler breeders should vaccinate twice to provide protection to chicks (live or killed vaccine).

Mycoplasma Gallisepticum (MG): Live MG vaccine is most effective if it is given to MG negative chickens before they become infected by the field strain. It provides less protection if given before 4 weeks. Other programmes are available for breeders.

Infectious Laryngotracheitis (ILT): ILT can be prevented by good isolation and sanitation procedures. The vaccine virus is live and may revert to pathogenic virus and is shed for life by vaccinated birds (as is field virus by survivors of outbreaks). In endemic areas breeders and egg type chickens should be eye drop vaccinated when moved to the laying house. Vaccinate 2 weeks before infection is likely to occur if growing birds are at risk. In some places broilers are vaccinated in the drinking water at 3 weeks.

Fowl pox: In endemic areas pox may be spread by mosquitos and cannot be prevented by isolation. Live pigeon pox (or other types) vaccine may be given by wing web stab at 8 to 12 weeks. Turkeys may also require protection.

Egg Drop Syndrome (EDS, Adenovirus Group III): In countries where EDS occurs layer and breeder flocks may be vaccinated at about 16 weeks with killed oil-adjuvant bacterin.

Adenovirus Group I (Inclusion body hepatitis, IBH): There are 12 or more serotypes of adenovirus Group I. Subtype 4 and 8 may cause disease without previous immunosuppression. Live vaccines may have to be prepared to prevent these infections.

Fowl Cholera: On farms or areas where fowl cholera occurs, birds may be vaccinated with killed or live vaccine. There are many strains of Pasteurella multocida and autogenous bacterins may be required for protection.

Infectious Coryza: As with fowl cholera. Live vaccine is not available.

Coccidiosis: Chickens or turkeys raised on litter or outdoors should be protected by medication until they develop immunity or by oral vaccination at day 1.

Turkeys may be vaccinated against Adenovirus Group II (which causes hemorrhagic enteritis (HE) with live vaccine and against erysipelas with bacterin.

Prevention by Medication

- Preventive medication Some diseases such as coccidiosis, necrotic enteritis and enterohepatitis can be prevented by medication. Preventive medication is most useful when protection is only required for a limited time as in broiler chickens or when immunity does not develop such as in necrotic enteritis.

Coccidiosis: Anticoccidial medication is very widely used in broilers and is generally used until the birds are ready for processing. Resistant strains of coccidia may develop and anticoccidials may have to be switched regularly. Immunity may or may not develop and anticoccidials that allow immunity are best.

Necrotic enteritis (NE) (Clostridium perfringens type A): Low level antibiotic in the feed is used to prevent NE and in Quail ulcerative enteritis (UE) (Clostridium colinum). The bacteria must be sensitive to the drugs being used and if predisposing factors such as coccidia infection are present they must also be corrected.

- *Therapeutic Medication:* Therapeutic medication can be considered preventive when it is used to control the spread of serious infectious diseases such as coryza or cholera. Treatment of coryza should not start until 1/3 of the flock is infected or the disease may recur following treatment.

  Some bacteria quickly develop resistance to medicine and different drugs may have to be used for control. Virus infections do not usually respond to medicine, although the medicine may prevent bacterial infection secondary to disease caused by virus.

  Medicines given by injection should not be given into the abdomen or leg. They can be given under the skin of the back or into the muscle of the breast.

Medicine given in the drinking water can be poured into the drinkers. In an automatic system they can be mixed in a large container and run into the system by gravity or a pump. Medicine can be added to a pressure system with a proportioner. Birds drink more water in hot weather. The level of medicine must match daily consumption and should be reduced in hot weather. It could be given for just 8 to 16 hours a day.

Medicine is often added to the feed at the feed plant. Most preventive medicine is used this way. In a disease outbreak medicine can be added to the water until medicated feed is available. Growing birds drink about 3 times as much water compared to the amount of feed they eat. Adults about 2 times as much (by weight).

Diagnosis of Poultry Disease

History. A good history will often provide clues that will help solve a problem. Get information on the type of bird, age, feed and water source and consumption rate, growth, production, morbidity and mortality, the description of the case, previous problems, vaccination programme, medicine being used, etc.

The problems may relate to management, environmental factors, and stress rather than to infection so examine the yard and housing conditions. Is the ventilation adequate? Are ammonia fumes a problem? Is it too hot or too cold? Is the litter wet or is it too dry and dusty? Is the pen too light or too dark? Are there sufficient hours of light for best production? Is the nest area darkened? Are the roosts too high? Do the birds appear comfortable? Chickens can talk and the sounds they make can indicate comfort, hunger, pain, panic, or disease.

Examination of Live Birds. Check the general appearance of the individual or group and try to determine which organ or system is involved in the illness. Note any signs or lesions that might point to a diagnosis, such as small size with poor feathering that suggests infectious stunting (malabsorption syndrome). If the birds show lameness or paralysis, is the lesion in the nervous system, bones, joints, muscles or skin? Some conditions, particularly those affecting locomotion, are easier to diagnose in live birds. Botulism which produces neck paralysis in chickens (leg and wing paralysis are more obvious in turkeys, ducks and pheasants) is an example.

Examine the skin of the head, body and legs for lice and mites, injury (particularly cannibalism), blood, mottling, swellings, anemia, cyanosis, or dermatitis. Listen for unusual breathing sounds (snicking, gurgling) and look for gasping or head-shaking that might indicate respiratory distress. Mouth-breathing (panting) is normal in chickens in hot weather. Exudate from nostrils and eyes and dirty feathers also suggest respiratory infection, or if just the eye, ammonia burn, ILTor eyeworm. Examine the droppings for evidence of diarrhea or other abnormalities. Take a blood sample for hematology or serology if indicated.

Necropsy. If a postmortem examination is to be carried out, birds that are representative of the problem in the flock must be selected. If birds have died, both sick and dead birds should be opened. Cull birds will not provide the answer. If the problem is a drop in production, try to find birds that look like they have recently stopped laying. It is important to do both an external and internal examination and to follow a specific routine to avoid missing important lesions.

- CRD - chronic respiratory disease
- IB(V) - infectious bronchitis (virus)
- ND - Newcastle disease
- AI - avian influenza
- CA(V) - chicken anemia (virus)
- EDS - egg drop syndrome
- IBH - inclusion body hepatitis
- ILT - infectious laryngotracheitis
- IBD - infectious bursal disease.

Live birds may be killed by cervical dislocation except when anemia or respiratory disease is suspected.

Examine the head, including the eyes, ears, nostrils, comb, wattles, mouth and beak.

Infected eyes and conjunctivitis are seen in CRD, sinusitis, Bordetella and Orthobacterium infection, infectious coryza and in ILT, IB and ND.

Keratoconjunctivitis with central corneal ulcers suggest ammonia burn. Swollen sinuses are seen in infectious coryza (Parahaemophilus infection) in chickens, sinusitis (mycoplasma), cryptosporidia and Bordetella infection in turkeys and pheasants. They may also be seen as the result of other infections such as influenza.

Chronic fowl cholera causes swollen wattles and sinuses in adult chickens. Fowl pox causes scabs on the comb, eyelid, and wattle, but must be differentiated from injury. Pox also occurs in turkey, pigeons, doves, canaries and other birds. If the bird has had its beak trimmed, check for proper healing, overgrowth of the lower beak or over-trimming (cut too short).

Injury on the head, neck or breast or back may indicate predators. Dermatitis and scabby or crusty lesions around the mouth and eyes suggest vitamin B deficiency.

Cut across the upper beak and back into the sinuses; then through the mandible and down the esophagus into the crop.

White plaques in the mouth, esophagus, or crop may be caused by capillaria worms, yeast infection (candidiasis, moniliasis), or possibly trichomoniasis or vitamin A deficiency, but most frequently by fowl pox (wet form). White plaques beside the tongue in the mouth are common in hens and turkey breeders and may be caused by mycotoxin or low humidity. Vomitoxin may produce similar lesions in young chickens and turkeys.

If the crop is enlarged and full, it may be an impacted or a sour crop (pendulous crop). The problem may be caused by excess water intake, defects in the crop itself, partial blockage of the proventriculus or gizzard or Marek's disease. Necrosis of the crop in wild birds is caused by Salmonella infection.

Examine the soft palate and larynx and cut down the trachea.

Wet fowl pox lesions are seen on the roof of the mouth, pharynx and larynx.

Granulation, congestion and mucus in the trachea are seen in CRD, IB, coryza, ILT, Orthobacterium, Bordetella, and *E. coli* infection, or with dust and ammonia fumes.

Hemorrhage and blood clots may occur in ILT and in ND or coryza and may cause severe gasping.

Gapes (gape worms) in pheasants, quail, and other birds is caused by Syngamus infection which also causes gasping. Cyanthastoma cause similar infection in water fowl and tracheal flukes may be found in waterfowl. Tracheal and air sac mites occur in cage birds.

Check the abdomen for lice and mites, and the vent for injury and prolapse of the oviduct.

Cut down between the abdomen and the legs and dislocate the hip joints. Peel the skin off the abdomen and breast. Tightly adhering skin and dark tissues indicate dehydration. Remove the breast carefully by cutting through the abdominal muscles, ribs and coracoids.

At this point, examine the thoracic and abdominal organs, paying particular attention to the air sacs, lungs, and liver. Carefully raise the gizzard and intestines and examine the abdominal air sacs, peritoneum, spleen and the ovary in laying birds.

The air sacs are cloudy in respiratory disease such as *E. coli* infection, Mycoplasma, Aspergillus, IB or early CRD in chickens. If there is also fibrin on the liver and in the pericardial sac, suspect *E. coli* or Salmonella infection. These lesions in turkeys and other birds might also be due to Chlamydia, or in ducks Anatipestifer infection.

Pneumonia in turkeys is caused by fowl cholera (Pasteurella multocida infection), and the lungs are quite solid. Aspergillosis, ND, AI, E. coli and Ornithobacterium infection can also cause pneumonia.

Peritonitis in layers is usually "egg peritonitis" caused by *E. coli* infection from the oviduct, although acute fowl cholera also causes peritonitis.

White crystals on the heart, liver and other tissues and organs are uric acid crystals (visceral gout) and are caused by hyperuricemia from urate nephrosis secondary to water deprivation, urolithiasis or other kidney disease.

Large blood clots in the abdomen or hemorrhage and hematoma on the liver are the result of trauma or fatty liver hemorrhage syndrome in chickens, or ruptured aorta in turkeys.

Tumors in or on the organs may be Marek's disease, lymphoid leukosis, or other varieties of tumors. Multiple small tumors on the organs and peritoneum in adult hens are metastasis from a carcinoma of the oviduct. This may result in ascites.

Ascites may also result from heart or liver disease or from ingestion of some toxic material. Right ventricular failure occurs most frequently in meat-type chickens after 4 weeks. It occurs secondary to pulmonary hypertension causing right ventricular hypertrophy and valvular insufficiency.

Focal white lesions on the organs may be tumors or they may be tuberculosis or coli-granuloma or, if just on the liver, blackhead. Turkeys, pheasants, and peacocks are all quite susceptible to blackhead; chickens are less susceptible. There are prominent cecal cores in blackhead (Histomoniasis) and occasionally in salmonella infection or coccidiosis.

In chicks, turkeys and waterfowl, examine the lungs and air sac for yellow-white foci or plaques caused by brooder pneumonia (Aspergillosis). Gasping in young birds is a sign of tracheal or bronchial epithelial injury, or obstruction and can be caused by irritating fumes or by infection (often IB plus *E. coli* or Aspergillosis).

In young chicks, look for an infected navel and for yolk-sac infection or peritonitis (mushy chick, omphalitis) in which the abdomen is swollen, wet, and discoloured, and the yolk-sac is infected (due to *E. coli*, Salmonella, Staph., Proteus, etc.).

Young birds also die because they don't start to eat (starve-outs) or drink (dehydration). Deaths occur mainly at 3, 4, and 5 days. This may be a management problem (chilling, feed and water not available, etc.) or the chicks/poults may be weak or defective when hatched.

Broiler chickens that die suddenly from sudden death syndrome (dead in good condition) or from heat stroke, or suffocation (piling-up) have congested, edematous lungs, a full digestive tract and congested mottled breast muscle.

Examine the circulatory and immune systems, the heart, pericardial sac, blood, spleen, bursa, thymus, and lymphoid tissue of the thigh and intestine.

Birds that die from anemia are pale and the blood is watery. With CAV the thymus is small and bone marrow may be pale. Birds that bleed to death (pick-outs, ruptured fatty liver, acute cecal coccidiosis, ruptured aorta, hemorrhagic enteritis in turkeys, etc.) are also pale.

Sulfa poisoning produces anemia and widespread hemorrhage in the tissues. Chicken anemia virus (CAV) produces similar lesions and is the agent responsible for infectious anemia associated with IBH. Lead poisoning may also cause anemia.

To identify anemia from parasites in blood cells (Plasmodium or leukocytozoon in ducks, turkeys and chickens), blood from a live, sick bird must be examined.

IBD (Gumboro disease) is caused by a virus that damages the bursa causing illness in 2-4 week old chickens or in younger chickens destroys part of the immune system, making birds more susceptible to other infections.

Swelling, congestion and hemorrhage with or without focal necrosis in the spleen, liver and lymphoid tissue suggest septicemia (fowl cholera, fowl typhoid, streptococcosis, or erysipelas) or viremia (ND and duck virus enteritis also affect the lymphoid tissue in the intestine).

Marek's disease and lymphoid leukosis produce tumors in lymphoid tissue and organs except for the bursa which is mainly affected by lymphoid leukosis (occasional Marek's lesions may occur in the stroma of the bursa).

Skin leukosis is Marek's disease virus causing viral dermatitis in the feather follicles. At processing this can be confused with scabby hip or other causes of dermatitis. Marek's disease can also cause lymphoid neoplasia in the skin.

Cut through the proventriculus and remove the digestive tract and liver. Open the proventriculus, gizzard, and small and large intestines to the cloaca. Check the cloaca carefully for evidence of picking injury.

If hens are not properly beak-trimmed or are too fat, mortality from "pick-out" is common. The whole intestine may be picked out through the cloaca. Prolapse of the vagina (and cloaca) (blow-out) may occur from excess fat or straining, secondary to injury or inflammation. Injury is common in flocks that produce large eggs before they become mature.

Examine the digestive tract for lesions and the various kinds of enteritis (hemorrhagic, necrotic, ulcerative, etc.), parasites (tetrameres, roundworms, capillary worms, tapeworms, cecal worms, and coccidia), and gastrointestinal accidents. A large proventriculus in broilers is from lack of fibre in the diet resulting in poor development of the gizzard. A thickened proventriculus sometimes with ulcers or hemorrhage may be Marek's disease or infectious proventriculitis.

Green staining of the digestive tract is just bile and indicates that the bird is not eating. The liver and spleen may be small (if the bird is thin) and the gallbladder full. Salmonella pullorum (pullorum disease, bacillary white diarrhea) causes enteritis, diarrhea and death in chicks. It has been eradicated in many countries.

Check the ceca, intestine and liver for lesions of blackhead, TB, coccidiosis or tumor, the liver for other varieties of bacterial, viral or protozoal hepatitis, cholangiohepatitis, etc., and the pancreas for tumors. A large liver may be lymphoid leukosis or Marek's disease, or it may be caused by bacterial hepatitis (*E. coli*, campylobacter) or fowl typhoid (Salmonella gallinarum).

Hepatitis with necrosis and hemorrhage that looks like bacterial (vibrionic) hepatitis may be immune damage to veins (vasculitis) from amyloid.

A large, yellow liver may be normal fat storage in a laying bird (estrogen stimulation) but layers die from fatty liver hemorrhage syndrome. This occurs when the liver ruptures because it has become fragile due to excess fat and free radical damage from fat.

A yellow or hemorrhagic liver particularly with focal necrosis may be viral hepatitis (IBH) which is seen in broilers, pigeons, raptors, owls and psittacine (but IBH does not cause infectious anemia).

Examine the testes or ovary and open the oviduct. Shrinking ova indicate illness of 2-7 days' duration or one from which the bird may be recovering. Small, sac-like ova indicate that the bird has been out of lay for a week or more and may be in a molt.

Semi-solid (cooked) ova indicate bacterial infection (such as salmonella).

An impacted oviduct may be secondary to vent-picking, egg-material left in the oviduct, or the bird may be egg-bound. Infection (Mycoplasma, IB, *E. coli*) can cause salpingitis as well.

A large or small fluid-filled cyst in the right abdomen beside the cloaca is the cystic remnant of the right oviduct (may be normal, but increased by IBV).

A drop in production may be related to clinical or subclinical disease (EDS, IB, MG, AE, AI, ND, etc.) or management faults (lack of light, temperature change, lack of water) or nutritional problems, etc.

Deformed shells suggest management faults, superovulation, EDS, IB and soft shells (higher than 1-2%) calcium or vitamin D3 deficiency.

A normal-appearing dead bird with an egg in the shell gland, or just laid has likely died from acute hypocalcemia. These birds often have fragile bones (particulary the femur), lack of medullary bone and rib infolding (osteoporosis, osteopenia, osteomalacia).

Hard (fibrotic) or swollen testes indicate bacterial infection (salmonella).

Examine the kidneys and ureters.

Swollen, pale or white spotted kidneys indicate hyperuricemia from urate nephrosis and may be due to lack of water or other kidney disease. Swollen. pale kidneys are also seen in IBH, IBD and fatty liver and kidney disease. Ureters plugged with hard stony material (urolithiasis) indicates a previous low phosphorus diet.

Swollen kidneys and nephritis may be due to IBV (nephrotrophic strain) or *E. coli* infection and usually causes death from dehydration. Newcastle's disease causes lympholytic foci in the kidney in pigeons.

Examine the skin, integument, muscles, bones and joints.

Emaciation, along with small organs, suggests malnutrition, stunting syndrome, beak injury (poor trimming), peck order (psychological) problems, chronic disease (coccidiosis), bumblefoot or other lameness, or chronic poisoning (lead, insecticide, etc.).

Muscular degeneration due to vitamin E-selenium deficiency can cause lameness, particularly in ducks. Ionophore toxicity causes muscle damage and paralysis in turkeys. Granulomas in the breast muscle are usually a vaccine reaction.

Sarcosporidial cysts produce small, white lesions in the muscle of water fowl.

Examine bones and joints for abnormality and deformity. Angular bone (valgus-varus) deformity of the intertarsal joint is caused by lateral or medial bending of the tibio-tarsal and metatarsal bones and is a common problem in meat-type poultry. It has a variety of possible causes (nutritional, rapid growth, management, etc.). Tibial dyschondroplasia causes backward bending or fracture of the top of the tibia. Slowing growth in young birds will help prevent leg deformity.

Other types of hock and stifle lameness and ruptured tendons are frequent in heavy roaster and turkeys and may be mainly due to injury as the result of heavy weight and fast growth, but some respond to added selenium or B vitamins.

Check for poor bone-breaking strength (osteoporosis) and, in young birds for rubbery bones, soft beaks, and beaded ribs which indicate calcium or Vit. D3 deficiency (rickets). Cage layer fatigue because of fragile bones may be due to phosphorus deficiency. Calcium and Vit. D3 deficiency also cause fragile bones (osteoporosis) in adults, but the most common cause is continuous high production.

Infectious stunting syndrome (malabsorption syndrome, fragile bones, osteoporosis) in young broilers is caused by intestinal damage from viral infection in young chicks. The chickens are small, poorly feathered and there is poorly digested food in the lower intestine. Osteomyelitis also causes fragile bones.

Curly-toe paralysis in young birds may be riboflavin deficiency, but in older birds and turkeys may be genetic or due to lack of roosts.

Cracked feet and foot dermatitis may be pantothenic acid or biotin deficiency, but scaly leg is caused by mites.

Footpad dermatitis and hock lesion are often caused by poor litter conditions. Toe injury in young birds may be cannibalism or mechanical injury. Arthritis in the feet or hocks or other joints suggests infectious synovitis (Mycoplasma synoviae) or *E. coli* or Staph. infection often with osteomyelitis.

Infection in the wing joints in pigeons is usually due to salmonella.

In broilers, roaster and broiler breeders viral arthritis (reovirus infection) may cause lameness or ruptured tendons. Ruptured tendons are usually caused by rapid growth and large body size (heavy weight) and there is usually thickening of the tendon above the hock.

If growing birds are lame, and there is no evidence of infection or rickets, cut into the proximal tibia and look for necrosis caused by osteomyelitis or dyschondroplasia and split the spine at T4 to look for spondylolisthesis (kinky-back), a plug of cartilage impinging on the cord or consider Marek's disease.

Necrotic dermatitis is caused either by staph. or clostridium infection and is associated with immunosuppression (usually by CAV or IBD). Scabby-hip is usually from overcrowding, poor litter conditions, poor feathering or scratches. Dermatitis or granulation in the neck may be a vaccine reaction, or contaminated vaccine.

Disturbances of the nervous system may cause incoordination, staggering, paralysis, walking backwards (with wings flapping for balance), tremors, stargasing, and other odd behaviour.

In case of lameness, paralysis, or incoordination, examine the sciatic nerves, spinal cord, and brain. Histologic examination will be required for diagnosis.

ND may produce CNS disturbances in pigeons as well as nervous, respiratory, intestinal and reproductive lesions in chickens, pheasants, turkeys, and wild and cage birds of all ages.

Range paralysis is a form of Marek's disease affecting the peripheral and central nervous system causing lameness, incoordination and paralysis. Marek's can also cause botulism-like paralysis and transient paralysis.

Vitamin E deficiency (avian encephalomalacia) causes soft, dark areas of hemorrhage in the cerebellum which may be visible grossly. B vitamin deficiency (thiamine and other) may also cause nervous signs.

Avian encephalomyelitis (AE) (epidemic tremor) affects birds up to 3-4 weeks old from non-immune parents.

Arsenilic acid and other feed additives and toxins (botulism, lead) may cause CNS disturbances, while others like the ionophores and coffee weed seeds (Cassia) cause nerve or muscle damage. Bacterial infection (pasteurella, pseudomonas, salmonella, staph. etc.) and fungi (aspergillus, etc.) also cause meningo-encephalitis, occasionally in outbreak proportions.

Hatching Eggs and Incubation

The Breeder Flock

Healthy chicks come from good breeders. Breeder nutrition, uniformity and the vaccination programme are critical.

Breeder Flock Health, Age and Nutrition

A healthy breeder flock, females with a good rate of lay mated to vigorous males, will usually produce highly fertile eggs that hatch well under good incubation practices. Hatching eggs from new breeder flocks (just starting to lay) do not hatch as well as eggs from those that have been laying for three to four months. Hatchability of eggs from old breeders nearing the end of their laying period also declines.

A hatching ration should be fed to breeders for about three weeks prior to saving eggs for incubation. The extra vitamins and minerals that it contains, as opposed to a laying ration, are essential for good hatchability. Sometimes, poor hatching results occur when a lower priced laying ration is used. The nutritional requirement of the breeder hen must be adequate for optimum hatching and good chick quality.

Clean Eggs are Important

Breeder house sanitation, nest box sanitation and hatching egg care are important to the production of healthy chicks. Nest material must be clean and dry and free of contamination. It should be cleaned daily (remove dirty material and droppings) and replenished or changed weekly. Para-formaldehyde pellets (20-25 g per nest, per month) will improve nest-box sanitation.

Breeder farms require a clean area where eggs can be cleaned, selected for quality and placed large end up on hatching trays. The temperature of this room should not be less than 21BC. If eggs are stored on the farm, temperature and humidity must be controlled.

Selection and Care of Hatching Eggs

Set only clean eggs. If eggs are soiled, it is preferable to clean them with sandpaper & egg brush. If hatching eggs are washed, the temperature of the water must be 43 to 44BC. The eggs should be spray washed with detergent, sanitised and fan-dried. Improperly washed eggs may rot or explode during the incubation period. Dirty eggs can be reduced if the eggs are gathered four or five times daily from nests supplied with fresh, clean nest material and if the litter in the breeding pens is kept dry so the hens feet are clean. Do not set floor eggs. Collect eggs with clean hands onto clean flats.

Select hatching eggs that are uniform in size (recommended minimum 52g for meat-type), shape and colour, with good sound shells. Do not set malformed, porous-shelled, doubled-yolked eggs or eggs with cracks. Shells that have a mottled appearance upon candling are not considered to have poor shell quality and can usually be set with good hatching results.

Since smaller eggs hatch in less time than large eggs it is recommended that eggs be separated by size and that small eggs be put in the incubator 8-18 hours after the larger eggs are set.

The hatching percentage will be the highest if eggs are held at a temperature of 16 to 17BC for not more than one week before setting. Higher temperatures initiate embryo growth. Storage temperature should be reduced to 13BC if eggs are being held for two weeks or longer. Eggs may sweat when moved to warm, humid areas. This allows bacteria to penetrate the shell. Turning eggs during the holding period is not beneficial. It has been shown that eggs held for more than 2 weeks hatched better when stored small end up (contrary to the accepted traditional large end-up postion). Relative humidity should be maintained at approximately 80% in the egg holding room. Higher humidity encourages mold growth. Prior to placing eggs in the incubator, they may be removed from the egg storage room and warmed to room temperature for approximately 6 hours.

The Incubator Room

In the selection of a successful incubator room, factors such as heating, humidity, ventilation, and sanitation should all be considered. Optimum results can be expected if the temperature can be maintained at between 24 and 27BC, with uniform humidity below the level that is required in the incubator. Tropical climates (heat & humidity) make it difficult to maintain good incubator room conditions. Good ventilation and a constant supply of oxygen to remove excess carbon dioxide from the environment surrounding the incubating eggs is necessary for the developing embryo. High altitude reduces available oxygen. An incubator should not be placed near an outside wall or window in cold climates or in direct sunlight.

Incubators

Incubators are the most important equipment in the hatchery process. Many kinds of incubators are manufactured; however, the general principles of all modern machines for commercial hatchery production are the same. Incubator setting capacity ranges from approximately 14,000 to 100,000 eggs.

During incubation, the hatching eggs are set vertically, with the large ends up in trays or flats in a setter and turned mechanically until about three days prior to hatching (setting period). The eggs are then transferred to a hatcher (hatching period) in a horizontal position and not turned during the hatching process. Both setters and hatchers have forced-draft air circulation, automatic temperature, humidity and cooling controls.

For small backyard poultry operators, there are small, still-air machines. Capacity of these incubators varies from 12 to 200 eggs. The eggs are set in a horizontal position and are usually turned manually. This type of incubator may be used for the entire incubation period for any kind of eggs. The source of heat is a thermostatically controlled heating element or light bulb. Humidity is supplied in most cases by water in a pan below the eggs, and ventilation controlled by small air vents or holes. Most still-air machines have transparent plastic domes through which the incubation process can be observed.

Table 1: Incubation Time for some Common Avian Species

Chicken	- 21 days
Turkey	- 28 days
Japanese quail	- 17 days
Guinea fowl	- 26 days
Pheasant, Partridge	- 24 days
Duck	- 28 days
Muscovy duck	- 35 days
Goose	- 28-32 days
Egyptian goose	- 35 days

The Principal Factors in Incubation

The five main factors affecting incubation, listed in order of importance are: temperature, humidity, ventilation, position, and turning of the eggs.

Some factors to consider before purchasing an incubator:

1. Any incubator will produce better results when operating at capacity. Egg capacity is a factor;
2. How long the incubator will last and cost of all replacement parts;
3. Service and availability of parts;
4. Accuracy of controls, to keep temperature and humidity fluctuations to a minimum;
5. The amount of labour involved in operating the machine and carrying out a thorough sanitation programme;
6. The guarantee.

Temperature

Hatching eggs may be warmed to a temperature of 25 to 30BC, prior to setting. The normal development of the embryo is dependent on the heat being held within a very narrow range in the incubator. In small still-air incubators, the temperature of the upper surface of the egg is higher than on the lower surface, while in large incubators, the air movement maintains the same temperature over the entire surface. For this reason, a still-air incubator must be operated at a higher temperature than a forced-draft incubator.

In small still-air incubators, a constant temperature of 39BC is considered satisfactory to produce good hatching results. The temperature may vary between 37.5 and 39.5BC without hurting the embryos as long as the temperature does not remain at either extreme. These readings should be taken with the bulb of the thermometer level with the upper surface of the eggs, but not in contact with the egg shell. A standing thermometer will give a more reliable reading than a hanging thermometer and the thermometer must be accurate. The temperature may rise one degree at hatching time without causing any reduction in hatch percentage.

In large incubators, the temperature, humidity, and speed of air movement are very closely dependent on each other, and since air speed varies in different incubators, it is impossible to state an exact operating temperature for all large machines, but generally it is around 37.5BC for a setter and 37BC for a hatcher. Follow the manufacturer's instructions closely with regard to temperature and ensure that instructions are for the model in use. High temperatures even for a very short period of time during any part of the incubation period will cause more harm than low temperatures.

Numerous factors may contribute to high or low incubator temperatures.

- high or low room temperatures or floor temperature under the incubator affect the operating temperature of both large and small incubators. A large incubator placed near an outside wall may have one section operating at a lower or higher temperature than the other. These problems and those listed below will result in one group of eggs hatching earlier or later than the others, increased embryo mortality or cause leg deformity in chicks & poults.
- floor temperature variation because of drains causing cold or heat under the incubator.
- an incubator thermometer not reading correctly.
- the failure of automatic equipment such as thermostats, cooling coils or automatic dampers.
- improper air circulation leading in turn to a rise in temperature in part of the incubator, because of: a decrease in fan speed, usually caused by low voltage or a slipping fan belt; incorrect spacing of filled egg trays in a partially filled incubator.
- overloading an incubator for any one setting of eggs.

Humidity

During the incubation or setting period, eggs should lose 11 to 12% of their weight (another 3 to 4% in the hatcher, after day 18), due mainly to a loss of moisture. The amount of moisture (humidity) in the incubator controls the rate of evaporation from the egg. The evaporation rate is also related to temperature, air speed, shell thickness, and size of eggs; the smaller the eggs, the greater percentage of moisture loss. Too great a moisture loss from the egg in the early stage of incubation will cause the embryo to adhere to the shell, causing death. Insufficient evaporation may cause death from lack of oxygen because of a small air cell, since just prior to pipping the shell, the embryo pips into the air cell and starts to breath air. The best guides to the correct amount of humidity in an incubator is the weight loss and the size and enlargement of the air cell during incubation, or the position at which the chick pips the shell. The degree of enlargement of the air cell should be determined by candling several eggs and estimating the averagee evaporation.

The amount of moisture in an incubator may be referred to as "relative humidity", which is a percentage of the moisture in the air at any given temperature. This can be measured by a wet-bulb thermometer.

A wet-bulb reading is based on air movement, and for this reason, a wet-bulb thermometer cannot be used to determine the amount of humidity in a still-air incubator. The relative humidity for small incubators is usually stated as 60%. Other inexpensive but less accurate equipment is available to measure relative humidity. A level water pan on the floor of the incubator under the eggs during the entire incubation period will generally supply sufficient moisture to give good hatching results, provided the moisture content of the air in the environment is neither too high or too low.

Percent egg weight loss during the setting period can act as a guide to the correct humidity. This can be done by weighing some eggs at the start of incubation and the same eggs again on the 7th and 18th day, and calculating the loss that occurred (write the weight on the shell with a pencil). Chicken eggs should lose approximately 4.5% and 11.5% during the first 7 and the first 18 days of incubation respectively.

Eggshells thicker than 0.34 mm are too thick and humidity should be reduced to increase moisture loss. Below 0.31 mm is too thin. Normal would be 0.33 to 0.34 mm at the beginning of production. Shell thickness decreases with age of breeder hen with 0.31 mm normal at 60 weeks.

In large incubators, the temperature, air speed and humidity are intimately related, and the manufacturer's instructions should be followed closely. Better hatching results may be obtained if the temperature is lowered and the humidity raised at hatching time (only if separate hatching compartments are available). Lowering the air temperature will provide an additional increase in relative humidity to keep shell membranes moist during the hatching process.

Table 2: Percent Relative Humidity At Different Wet-bulb Readings

Wet-bulb Reading (BC)	Relative Humidity (%) at Air Temperature of	
	37.2BC %	37.8BC %
32.2	70	68
31.7	67	65
31.1	65	63
30.6	62	60
30.0	59	57
29.4	56	54
28.9	53	51
28.3	51	48
27.8	48	46
27.2	45	43
26.7	43	41

Incorrect humidity may be due to a number of factors, the more important of which are listed below:

1. High environmental humidity in tropical countries.

2. A wet-bulb thermometer reading incorrectly. Remove the wick to determine whether the thermometer is reading the same as the dry-bulb thermometer.

3. Dust and dirt on the wet-bulb wick. Change wicks often and use only distilled water in the reservoir.

4. In small incubators with humidity supplied by water pan evaporation, ensure that there is always an adequate amount of water in the pan, as the rate of evaporation is dependent on the humidity in the room. The lower the humidity in the room, the higher the evaporation and vice versa. During hatching fluff settles on the surface of the water producing a film that causes a reduction in the rate of water evaporation necessary for optimum hatchability. Clean the water pan daily and replace with clean, lukewarm water.

5. Ventilating an incubator to control temperature will lower humidity in a dry environment unless automatic humidifiers are being used in the incubator room.

6. If airflow in incubators is not uniform humidity may be different from bottom to top or side to side.

Ventilation

The free movement of oxygen, carbon dioxide and water vapour through the pores of the shell is important, since the developing embryo must be able to take in a constant supply of oxygen and release carbon dioxide and moisture. Oxygen content of 21% (present in air at sea level) and a carbon dioxide content not exceeding 0.5% in the air are considered optimum for good hatching results. Room temperature, room humidity, the number of eggs set, the period of incubation, and the air movement in the incubator all influence ventilation requirements. Ventilation problems are not the same in small incubators as they are in large incubators, where a large number of eggs are set in a very small space.

During the early part of the incubation period, ventilation in small incubators may be held to a minimum. However, during the hatching period additional ventilation must be supplied to reduce the carbon dioxide in the incubator. It is advisable not to increased ventilation until half of the hatch has been completed, since ventilating too soon will reduce the humidity. In large incubators, the manufacturer's directions should be followed, however, ventilating recommendations may not be applicable to every locality and every room condition.

If ventilation is used to control either temperature or humidity in the incubator, the control of the same factors in the incubator room are important. At a room temperature below 18 C, ventilating an incubator will reduce both temperature and humidity. In a room with high humidity, (tropical countries) the primary concern is to maintain the correct temperature.

The main ventilation consideration may be summarised as follows:

1. Ventilation is more important in large incubators than in small incubators.
2. The amount of ventilation required may be altered by atmospheric conditions.
3. Ventilation is very important in any incubator at hatching time. Insufficient ventilation may result in embryo or chick death.
4. Ventilation in excess of the recommended amount may be applied to reduce temperature or humidity.

5. The appearance of chicks panting in a hatcher at normal temperature is an indication of a rise in the carbon dioxide content of the hatcher air. Under such conditions chicks must breathe faster to obtain the required amount of oxygen and to eliminate the exceses carbon dioxide. If excessive panting occurs, increase the airflow in the hatcher.

Position and Turning of Eggs

In small incubators, the eggs are maintained in a horizontal position during the entire incubation period. In large incubators eggs should be placed in a vertical position, large end up, during the hatching period. In small incubators, the eggs are moved when turned, while in large incubators they remain in a stationary position on the incubator tray and the egg tray is turned through an angle of not less than 90 in opposite directions with each turning.

The objective is the same in both types of incubators; namely, to prevent the embryo from sticking to the shell membranes. Turning also ensures a complete contact of the embryonic membranes with the food material in the egg, especially in early stages of incubation.

In small incubators, the eggs should be turned at least four times daily. It is advisable to leave some space on the tray to allow for moving the eggs forward a 1/2 turn on one turn and back a 1/2 turn on the next, thus making sure that all the eggs move.

Eggs should not be turned in a complete circle, as this has a tendency to rupture the allantois sac with resultant embryonic mortality. Wash hands carefully before turning eggs to avoid bacterial contamination of the shell. In large incubators, the trays are usually turned hourly with all the egg trays moving at one time.

For good hatchability, eggs should be turned to a position at least 45 from vertical, then reversed in the opposite direction to a similar position. In the most recent models of incubators, eggs are turned through an arc of 150 and in a few models they are turned as far as 180. The introduction of these newer methods of turning eggs has been an important aspect in improving hatchability.

Eggs should not be turned in either large or small incubators during the hatching period. The greatest benefit from turning eggs is during the first week in incubation.

Other Factors Affecting Incubation

Egg Selection

Poor quality hatching eggs do not hatch as well as eggs of good quality. The term "quality" refers to the condition outside the shell, the condition of the shell itself and that of the contents. Eggs with inferior characteristics, as discussed in "Selection and Care of Hatching Eggs," should not be set.

Sanitation

Eggs used for hatching should be clean and stored in clean containers in a sanitary egg holding room. Eggs contaminated with bacterial organisms usually do not hatch well and this poor quality is reflected in the chicks that do hatch.

Egg Handling

Rough handling of hatching eggs before they are set will increase the number of dead embryos, with mortality occurring between the 4th and 13th day of incubation. Also, jarring eggs during incubation may result in the rupture of the egg shell membrane and thereby lower hatchability.

Large fluctuations in temperature and humidity during storage will have a major adverse affect on hatchability. Refer to "Selection and Care of Hatching Eggs" for proper egg storage procedure.

Toxicity

If the interior of an incubator is painted or varnished, or if the trays are varnished, the percentage of hatch will be reduced, possibly by as much as 25%. This adverse effect disappears in about 30 days, suggesting that the ill effect is eliminated by oxidation of the paint.

This problem may be overcome without any reduction in percentage of hatch if the incubator is fumigated with formaldehyde gas at the concentration recommended for proper hatchery fumigation. The gassing should be done as soon as the paint is dry and with the incubator operating at recommended temperature and humidity for incubating eggs.

Automatic Equipment

The addition of automatic equipment has eliminated many of the problems with incubators, however, such hazards may occur when automatic devices fail. During hatching all automatic devices, such as cooling coils, automatic dampers, tray turners, alarm bells, etc., should be checked at regular intervals. Electrical failure may require that automatic equipment be reset. After an incubator is purchased, it is important to know how every part operates.

Egg Candling

Candling chicken eggs on the 7th and 18th day of incubation, may be recommended for small poultry producers. Egg candling will detect infertiles and early dead germs. Therefore, problems within the hatching flock can be identified without waiting until the incubation period is completed.

Improper Fumigation

Closely follow instructions outlined in "Fumigation Procedure". Excessive and improper fumigation can result in high mortality in developing embryos.

Fumigation of Incubators

The killing of bacterial organisms by formaldehyde gas is based on the concentrations of the gas, exposure time, temperature, and humidity of the incubator. The chemicals potassium permanganate and formalin (which is 40% formaldehyde gas) have proven to be the most effective method of destroying bacterial organisms in the hatchery. To accomplish the proper release of the gas, one and one-half parts (by volume) formalin is added to one part (by weight) of potassium permanganate. This will release the formaldehyde as gas or fumigant. When the reaction is complete, a dry, brown powder will be left. If the residue is wet, not enough permanganate was used; if the residue is purple, too much permanganate was added.

Concentration

The recommended concentration for effective fumigation is 53 mL of formalin added to 36 g of potassium permanaganate per cubic metre of space to be fumigated, or 1 1/2 mL of formalin added to 1 g of potassium permanganate per cubic foot of space to be fumigated.

Caution: Never add the permanagante to the formalin. Heat is generated when the two chemicals are combined, and care should be taken. Formaldehyde gas is generated quickly. Do not allow the fumes to get into the eyes. Personnel should use a respirator or wear a mask to avoid unnecessary exposure. Ventilate the incubator room to remove fumes that escape from the incubator.

Time

It is not recommended to fumigate setters with hatching eggs in them, but if such treatment becomes necessary, embryos between 24 and 96 hours of age should not be exposed to the above concentration of formaldehyde. Hatching compartments should be fumigated after the eggs are transferred from the setter to the hatcher, again after the hatch has been taken off and before the refuse has been removed from the trays, and finally after the hatcher has been thoroughly cleaned. Do not fumigate chicks with this concentration of formaldehyde gas. Small still-air incubators should be fumigated after the chicks have been removed and prior to discarding the refuse from the tray and again after the incubator has been thoroughly cleaned.

Fumigation Procedure

1. Make sure the temperature and humidity of the incubators are at normal operating conditions.

2. Measure the inside volume of the machine in cubic feet or cubic metres (length x width x height).

3. Close the ventilators, but leave the fans on.

4. Weigh the required amount of potassium permanganate into a wide enamelware or earthenware vessel large enough to accommodate the boiling and splattering action experienced when the formalin is added. Place the vessel and the permanganate in the area to be fumigated; then add the formalin.

5. Close the door immediately and leave closed for 20 minutes.

6. After 20 minutes, open the ventilators.

7. Open the doors of the machine for five minutes, leaving the fan on to allow more of the formaldehyde gas to escape, or neutralise it with a 25% solution of ammonium hydroxide equal to one-half the amount of formalin used. The hydroxide should be thrown directly on the floor of the machine and the doors closed. The formaldehyde gas will quickly be neutralised.

Continuous Fumigation of Hatcher

The greatest increase in bacterial organisms occurs during the hatching period. These can be reduced, but not completely eliminated, by slow release of formalin in the hatcher during the last 48 hours of the hatching period.

For continuous fumigation to be effective, hatching eggs should also be fumigated at transfer time with the recommended concentration.

Place formalin in pan about 25 mm in depth, allowing 58 cm2 of pan for each cubic metre of hatcher space, or a pan about 1 in. deep, allowing 30 in2 of pan for approximately 1000 ft3 of hatcher space. Do not use permanganate. The pan should be placed in the open area of the hatcher in direct line with the airflow.

Place the pan of formalin in hatcher 48 hours prior to hatch completion. To overcome fluff deposited on the formalin, add more formalin about 24 hours before hatch is complete to increase evaporation.

If the fan stops (mechanical failure, etc.) and reduces the airflow, remove the pan of formalin immediately. If hatchers are only partially full, continuous fumigation is not recommended, as a reduction in airflow will create an excessive build up of formalin in the hatcher.

Effects of Fumigation

1. Properly carried out, fumigation should not affect hatchability.
2. Fumigation will only kill bacteria that are present on the surface of hatchery refuse. Fumigation will not kill bacteria inside unhatched or pipped eggs. It is important to dispose of hatchery refuse carefully to minimise hatchery contamination.
3. The hatchery room must be separate from the tray dumping room and from the chick processing area. Air flow and traffic must be controlled to prevent contamination of the chick processing and holding areas.
4. Efficient fumigation along with other sanitary measures should control navel infection (omphalitis).
5. Fumigation is not intended to replace a thorough cleaning programme.

Hatchery Design

Hatchery layout is important to a good sanitation programme. Arrange the hatchery so there is a one-way flow of material from the point where the hatching eggs are brought in to where the processed chicks go out.

Hatcheries should be designed so the flow of eggs, chicks and personnel does not spread contamination from one room to the next. Doors, including one-way doors, help stop cross contamination between rooms. Positive air pressure prevents contamination through an open door. Workers should change outer clothing and wash hands before moving from one work station to another.

Summary of Some Important Factors

1. Feed breeder flock hatching ration that is well fortified with essential nutrients.
2. Use healthy breeding stock.
3. Provide good egg-holding facilities.
4. Avoid holding eggs in storage for more than one week.
5. Prewarm eggs for 6 to 8 hours at incubator room temperature.
6. Set clean, good quality eggs. Delay setting small eggs (those more than 10% less than average) for 8 to 16 hours.
7. Maintain correct incubation temperature, humidity and ventilation. Make sure air intake does not draw contaminated air into the incubator.
8. Turn hatching eggs frequently.
9. Maintain incubator room temperature between 21 and 24BC with good ventilation and relatively high humidity.
10. Fumigate regularly.
11. Clean vaccinating and beak trimming equipment. Newly hatched chicks may pick up contamination and infection in the hatchery from vaccinating and beak trimming equipment. This equipment requires a very rigid sanitation schedule.
12. Practice strict sanitation; cleanliness is very important for successful hatching operation. Make sure belts, equipment and workers hands used to move eggs or newly hatched chicks are kept clean.

Animal Welfare Problems in Chickens and Other Poultry Caused by Single Trait Genetic Selection for Meat and Egg Production

Chickens have several serious welfare problems that come from bad genetics and can be fixed only with good genetics The biggest problem in many intensively raised animals is pushing the animal's biology for more and more production. Breeders choose the most productive animals — the fastest growing, the heaviest, the best egg layers, and so on — and selectively breed just those animals. Bad things always happen when an animal is overselected for any single trait. Nature will give you a nasty surprise.

Bone breakage is a very serious problem in both caged and cage-free hens because laying hens have been overselected for egg production. Commercially bred hens put all their calcium and minerals into forming eggshells, and their own bones become depleted. Their bones are so weak that in cage-free systems a hen can break her leg just jumping off her perch. The only way to solve this problem is for the industry to accept the fact that birds with strong bones will produce slightly fewer eggs.

Laying hens have other problems, too, especially feather pecking and cannibalism. Feather pecking is what it sounds like: one hen pecks at another hen's feathers or pulls a feather out all the way. Severe feather pecking can lead to cannibalism, with the victim hen being wounded and then killed by the hen doing the pecking. Even though a feather-pecking hen can kill her victim, feather pecking probably isn't driven by the RAGE system. We know this because of studies mixing unfamiliar hens together. Aggression goes up, but feather pecking and cannibalism don't. They aren't the same thing.

Instead, feather pecking is probably displaced or redirected SEEKING behaviour. It's a kind of foraging or exploration of another bird instead of the ground. We know this because of research showing that chickens housed on litter do a lot less of it. They peck at the litter on the ground, not at each other's feathers. The more active the bird and the more foraging behaviour it does naturally, the more likely it is to develop severe feather pecking. Both feather pecking and cannibalism are affected by genetics.

Some modern broiler chickens have genetic problems related to growth. I was shocked to learn at a chicken-breeding seminar that the broiler chicken has been so overselected for rapid growth that its bone physiology is totally abnormal. In normal bone development, the body first "erects" a scaffolding or frame of cartilage and then fills in the frame with minerals that harden into bone. After the bone has hardened, the cartilage dies off through programmed cell death. In broiler chickens, something goes wrong with the cartilage, so the bones don't have support while they're hardening and end up misshapen. I liken it to building a new basement wall and taking down the plywood concrete forms before the concrete has fully hardened. In some of the worst cases, a chicken's feet are rotated almost 90 degrees and the legs are twisted. These chickens are genetically lame. Several studies have shown that lame broilers will choose feed laced with painkiller over their regular feed, and a study of lame turkeys showed that they started moving around a lot more once they were on painkillers. The industry has created chickens that have chronic pain in order to get birds that grow at the far outer limits of what is biologically possible. When an animal's biological system is pushed to the point where the physiology is totally pathological, I get disgusted.

The other problem is that modern broiler chickens have been bred to have stupendous appetites so they'll grow super-fast and reach market weight as soon as possible. The trouble is that the breeder chickens, the parents of the broilers, have the same stupendous appetites as their chicks. If you let a broiler breeder chicken eat everything she wants, she will become obese, her fertility will decline, and her life will be shortened. These chickens have to be kept on a strict diet just to maintain normal weight. They act miserable, and many of them develop stereotypies. These birds have low welfare no matter what you do. If you let them eat all they want, they have bad welfare and if you don't let them eat all they want, they also have bad welfare. It's terrible. The industry is going to have to breed parent stock with smaller appetites. There's no other way to fix the problem.

Then there are other genetic problems that no one understands. One of the worst cases was the rapist roosters. I wrote about them in Animals in Translation. Fortunately, the broiler industry has made some genetic changes to correct these problems, although there's still a way to go. The rapist roosters violently attack hens and injure and even kill them. Before the 1990s there weren't any rapist roosters. They just suddenly appeared out of the blue. First it was just one strain of roosters that had become aggressive but within a couple of years almost all strains had developed the same behaviour. Nobody knows why.

The rapist roosters have two problems: They are hyperaggressive and they have stopped doing the courtship dance the hen needs to see before she will mate. They've lost the little piece of genetic code that makes them do the dance. When the hens don't see the courtship dance, they don't become sexually, which may make the roosters' aggression worse. An unreceptive hen would be a form of frustration because it is a restraint on the rooster's action. So the RAGE system would be activated to some degree.

When I wrote Animals in Translation it looked like the rapist roosters were a side effect of the industry's selective breeding programme to create chickens with bigger breasts for more white meat. But now researchers aren't sure what caused it, or whether the hyper-aggression and the bad courtship behaviour are the same problem or two different problems that happened at the same time. Industry breeding programmes are trade secrets. It's obvious the industry is selectively breeding for larger breast size because breast size is getting larger. But we don't know what other selective breeding programmes the industry might be using.

Ian Duncan has an interesting theory about what might have happened. Dr. Duncan points out that big-breasted male birds have trouble mating because their huge chests get in the way. Male turkeys have such big breasts now that they can't mate at all and the hens have to be artificially inseminated.

Dr. Duncan says that if the same thing is happening to male chickens, the broiler breeder industry may have misdiagnosed the problem. When broiler breeders see chickens with decreased fertility, they attribute the problem to low sex drive. It's possible that the breeders who created rapist roosters were actually trying to increase roosters' sex drive. If the breeders selected for higher libido they could have mistaken a little bit of aggressive behaviour towards the hen for higher sex drive and ended up breeding hyper-aggressive roosters that for some reason had also lost their courtship dance. We'll probably never know.

Today the aggressive rooster problem has been greatly reduced although it hasn't been eradicated.

Better Breeding Strategies Using Group Genetic Selection

Most of the time breeders deal with genetic problems by culling chickens that have the problems and mating the ones that don't. Another interesting approach is group selection. The researcher Bill Muir at Purdue University has shown that you can reduce feather pecking genetically by using a technique called group selection. With group selection, instead of picking certain individuals, you pick certain family groups to breed. Dr. Muir has done this by raising several "sire family" groups — groups of chickens related to each other through their father — and then selectively breeding the group that has the highest egg productivity and the lowest amount of feather pecking and cannibalism.

Group selection has a couple of advantages over individual selection. First, the fact that you're working with groups instead of individuals means you know something about the birds' behaviour in a group. When breeders select high-productivity, individually housed laying chickens to breed, they don't know whether they are feather peckers or not because they've never lived with other birds. Individually housed laying hens can't express the behaviour.

Second, when you have a group of genetically related chickens living together you also see how living in a group affects their behaviour and productivity. When you choose which group to breed, you're not just choosing one genetic strain of chickens over another, or two desirable behavioural traits (good egg laying and low feather pecking) over undesirable traits. You're choosing one way of relating to the environment over another.

This is important because most behaviour is affected by what's going on in the environment. A hard-wired behaviour like the courtship dance is always the same no matter what's going on in the environment. It's like a computer subroutine; once you turn it on it just runs. But everything else is affected by the environment. When breeders use group selection instead of individual selection, they're factoring in the way the selected group relates to its social and physical environment.

Today, only a handful of companies provide all of the commercial layers and broilers around the world, which has greatly narrowed the gene pool. This has created a risky situation because genetically similar animals are vulnerable to the same diseases. Sure enough, when the Australians phased out their homegrown broilers and imported American birds, they ended up with more disease problems.

This is why it's important to preserve the old breeds of animals and poultry. Keeping the classic breeds alive is the only way to preserve genetic diversity and to save animals that have valuable genetic traits breeders may want to breed back into commercial lines in the future. The meat from some of the old breeds is more tender and better quality than meat from animals bred for rapid growth, and the chickens are hardier, too. They perform better in pasture-based or organic farms. They are beautiful, unique animals that shouldn't be destroyed by commercial breeding. Fortunately, many of the older breeds of poultry and livestock are being raised by local farmers and sold in farmer's markets or to gourmet restaurants. If a serious disease ever kills commercial broilers or layers, the entire world will be thanking the small producers and hobbyists who have kept the old breeds of chickens from becoming extinct.

Reproductive Responses to Sel-Plex® Organic Selenium in Male and Female Broiler Breeders: Impact on Production Traits and Hatchability

Achieving success with broiler breeder management is like hitting a moving target. Modern broiler stocks have been reported to grow at 4.6 times the rate of a 1957 random-bred strain due to increased genetic potential (Havenstein et al., 2003a). The 6-fold improvement in carcass yield of 2001 stocks fed a 2001 diet compared to 1957 stocks fed a 1957 diet is 85-90% due to genetics, and 10-15% due to nutritional changes (Havenstein et al., 2003b). While broiler 42- day body weight is increasing each year, the 42-day target body weight for male and female broiler breeders has remained the same, or even decreased (Rustad and Robinson, 2002).

In 1979, Hubbard male and female breeders were approximately 50% of the 42-day broiler weight. In 2001, this percentage had decreased to 36.1 for males and 30.3 for females. In essence, the degree of feed restriction has continued to increase while there is increased competition for a reduced feed allocation.

Complications in managing this increased growth efficiency is further complicated by the development of 'yield' varieties, carrying increased amounts of breast muscle, often on a smaller carcass frame. As selection for broiler breeder egg production is not as heritable or profitable as selection for growth traits, there are continued increases in the growth potential while egg production is not emphasised. As a result, Whitehead (2000) indicates that geneticists continually compound the problem by breeding a bird that, if allowed to exist in its freely expressed adult state, is completely unfit for life.

Understanding the ovarian function of the chicken and its interaction with nutritional status, age, and strain is likely the most important issue affecting poultry breeding companies today. The process involves the conversion of genetic, environmental, and nutritional cues into a cascade of signals from the neuroendocrine system.

These signals must be integrated and responded to by the organs and tissues primarily involved in reproduction, which will in turn produce more signals for both local and distant activities. The resulting eggs produced are the net result of the bird's attempt to coordinate the demands its body and environment have placed on it. The ability of an embryo to survive the incubation process relies on a balance between hatchery management and breeder management. Specific feed ingredients, bird age, and flock management decisions can directly affect semen quality, the oviduct environment, and the egg environment. These factors combine to influence the potential of the egg to be fertile and ultimately to hatch.

Reproduction in the Broiler Breeder

The reproductive system of the laying hen is comprised of many organs. The list includes the hypothalamus, the anterior pituitary, the ovary, the oviduct, the liver and the skeletal system. Small follicle steroidogenesis, particularly estrogen production, is responsible for transforming a pullet into a hen. As plasma levels of estrogens increase, externally visible features include reddening and enlargement of the comb and wattles, a prenuptial feather molt (feather drop) and a widening of the pubic bones to permit egg passage. Internally, estrogen stimulates liver production of egg yolk lipids with a significant change in the colour and size of the liver. Finally, the oviduct enlarges and becomes a secretory organ for deposition of albumen.

The male focuses on quantity rather than quality when it comes to sperm production. The hen then must screen out unsuitable sperm in order to guarantee production of high quality chicks. Mature sperm spend the majority of their time in the oviduct. Following mating, the hen will store sperm in the highly specialised microenvironments of the sperm storage tubules (SST) are located in the vaginal region of the oviduct. Only about 1 to 2% of the originally inseminated sperm enter the SST (Bakst et al., 1994), where they exist in a near-dormant state. The survival of the sperm to the point of insemination depends on a combination of sperm quality and the ability of the hen to provide a safe environment for the sperm.

The hormone messages being relayed between the ovary and the hypothalamic and pituitary control centers are altered by feed intake. Besides affecting follicle formation, and reproductive control, feeding level can alter the viability of the embryo through changes to the egg and to the early maturation process.

Excess nutrients are diverted into liver lipids, excess ovarian follicle development, and as abdominal fatpad (Etches, 1996). It can be a vicious cycle, with obesity continuing to worsen as the rate of egg production remains low and/or goes into early decline due to excess feed intake. The ovaries of growth-selected strains are particularly sensitive to overfeeding during the sexual maturation process.

Body weight in broiler breeder hens has been reported to be negatively correlated with duration of fertility and fertile egg production. Ultimately, reduced chick production in overfed broiler breeders is the culmination of poor egg production combined with reduced fertility, hatch of fertile, and embryonic viability.

The female oviduct environment can be hostile to sperm despite their existence in the SST within the oviduct wall. Duration of fertility (measured by monitoring fertility in consecutive eggs) can be reduced under conditions of overfeeding. It is known that fewer sperm survive in some bird strains and when excess feed is used, but it is not clear how the surviving sperm are affected, and if the remaining sperm are of similar quality to the ones originally placed.

Factors Affecting Hatching Egg Quality

There are many factors that can affect the potential of the embryo to survive incubation and generate a quality chick. Some of these factors are out of our control, such as hen age, and others can be manipulated through management decisions, such as egg size and hatchery environment. It can be difficult to formulate diets to Optimise egg production, fertility, and hatchability as little is known about the nutritional requirements of the embryo.

Dietary vitamin levels are increased in the diet with the hope they will also be increased in the egg. Yet there can be adverse reactions with this type of approach, as some vitamins have antagonistic relationships with others. Furthermore, there can be stability issues for long-term storage, as well as for feed processing procedures. With current genetic stocks, if the chick hatches in a weakened state due to a vitamin or mineral deficiency, it is more likely to succumb to disease now than with previous stocks.

Growth-selected stocks have low immunoresponsiveness (Siegel et al., 1984) due to either inadvertent or intentional negative selection pressure combined with growth efficiency selection. The developing embryo is especially sensitive to vitamin deficiency, which will result in death, malformation or some other atypical response.

During embryo development, oxidative metabolism increases substantially over the incubation period and especially in the last few days before hatch. This normal respiration related to embryo growth results in the production of free radicals, which can cause tissue damage through lipid peroxidation, with polyunsaturated fatty acids being especially vulnerable. The chick has developed effective antioxidant pathways to prevent damage.

The primary defence mechanism is a group of three enzymes (superoxide dismutase, glutathione peroxidase, and catalase), which convert free radicals produced by cellular respiration into less harmful alcohols (Ursiny et al., 1997). A second level of defence are the natural antioxidants – vitamin E, carotenoids, ascorbic acid, and glutathione, which protect the developing chick (Surai, 1999). During the last week of incubation, fat-soluble antioxidants are moved into the liver and yolk sac membrane.

The major fat soluble antioxidant, vitamin E, moves from the yolk to the embryo tissue at this time. Ascorbic acid (vitamin C) is the major water-soluble antioxidant, and is produced in the yolk sac membrane before transport to tissues like the brain (Surai, 1999). This helps protect membrane lipids during the large metabolic effort of hatching. The third level of antioxidant defence is the generation of enzymes that rebuild damaged membranes.

7

Role of Dietary Selenium

Selenium is normally provided in the diet in the form of inorganic sodium selenite. An organic form can be provided (Sel-Plex®), which is selenium yeast.

Yeast, like plants, form selenoamino acids and other organic selenocompounds that exist in very reduced form in comparison to the highly oxidised inorganic selenium forms (selenite and selenate). Organic minerals are transported intact and retained better in target tissues or organs. Higher selenium in eggs reflects increased antioxidative properties of the egg during storage, therefore preserving the egg for incubation and potentially increasing hatchability.

Cantor (1997) and Paton et al. (2000b) found that eggs from Sel-Plex®-fed chickens were significantly higher in selenium than eggs from sodium selenitefed chickens. Organic selenium has a vitamin E-sparing action through its involvement in vitamin E retention in the plasma and through involvement with the primary enzymatic defence system of the embryo against lipid peroxidation. In fact, supplementing organic selenium to breeder diets has been shown to increase levels of other antioxidants (vitamin A, E and carotenoids) in the egg (Surai and Sparks, 2001).

The protective effects of organic selenium are especially apparent during the highly oxidative state of late incubation and the first few days after hatch.

Selenium is an integral part of the antioxidant enzyme glutathione peroxidase (GSH-Px) as well as a component of many other selenoenzymes. Oxygen metabolism produces free radicals, which have potentially toxic effects on all biological molecules (Surai, 2000). Glutathione peroxidase aids in the removal of oxidative compounds in the form of hydrogen peroxide and hydroperoxides from the cell (Burk, 1989). Buildup of these substances can impair cell membrane structure and function, and once the membrane is damaged decreased productivity and reproductive performance can result (Surai, 2000).

Selenoamino acids have been shown to have higher bioavailability than traditional inorganic sources commonly used for dietary supplementation. They are actively absorbed in the intestine compared to the passive absorption of inorganic selenium (Surai, 2002). Furthermore, selenomethionine and selenocysteine can be incorporated non-specifically into structural proteins (particularly muscle tissue) during protein synthesis.

Selenomethionine can be substituted for methionine during protein synthesis due to its similar structure (methionine contains a sulphur atom instead of a selenium atom). This critical difference between selenium sources allows the organic selenium compounds in Sel-Plex® to contribute to a selenium reserve to be available for prevention of lipid peroxidation (through GSH-Px) during stress conditions (Surai, 2002). In the broiler breeder, it also enables enhanced transfer of selenium from the hen to the embryo (Edens, 2002). Increased antioxidant uptake in the hen due to the maternal diet is linked to increased antioxidant concentrations in the developing chick (Surai et al., 1999).

The need for defence against oxidative damage is clear in the male, where antioxidant enzymes play a key role in maintaining the sperm cells (Surai et al., 1998). Sperm cells contain large amounts of polyunsaturated fatty acids, which allow them to maintain flexibility relating to motility (Surai, 2002). However, this means they are also a target for lipid peroxidation. Cellular integrity is maintained by GSH-Px, other selenoenzymes and vitamin E, which protect the cell membranes from oxidative damage (Flohe and Zimmermann, 1970).

Some recent research has demonstrated that the inclusion of selenium in poultry diets enhances sperm numbers, and using an organic source (Sel-Plex®) reduces production of defective sperm, thereby having a positive effect on the fertilizing potential of the male (Edens, 2002). Little information is available regarding the effect of dietary selenium source on the reproductive efficiency of laying hens.

Egg production and fertility decline with age. The decrease in hatchability and fertility associated with an increase in age might be due to the older hen's inability to hold sperm in the SST.

Furthermore, the sperm do not retain their viability as long in the SST of older hens, and are released in larger numbers from the SST (Bramwell et al., 1995). Quantification of fertility is determined by the occurrence rate of perivitelline holes caused by the sperm. In past research, killing the hen to obtain the newly ovulated ovum was the only way to determine sperm hole quantities. Bramwell et al. (1995) adapted the technique to use eggs for the determination of perivitelline sperm hole numbers.

This study examined the effects of selenium supplementation form and level on female reproductive performance and egg traits. Its intent was to determine the effects of organic (Sel-Plex®) and inorganic selenium supplementation in the laying hen diet on fertilization potential and egg traits.

Methods

We housed 75 hens in individual laying cages at 61 weeks of age. Three dietary treatments were imposed, varying in selenium source and level. Twenty-five hens were fed a control diet, 25 were fed a diet enriched with inorganic selenium in the form of sodium selenite and 25 were fed a diet enriched with organic selenium in the form of the Alltech product, Sel-Plex®. All diets contain 19% CP and 2875 kcal ME/kg. The control diet had a selenium inclusion rate of 0.1 mg Se/kg, whereas both enriched diets contained a total of 0.3 mg Se/kg with an added 0.2 ppm Se coming from the organic or inorganic selenium source. The diets were fed for a 3-week period prior to insemination to ensure tissue saturation of the new dietary selenium forms and concentrations.

Following the 3-week acclimation period, all hens were artificially inseminated with 50 µL of neat, pooled semen collected from a group of 22 broiler breeder males (116 million sperm/dose). Eggs were collected from 2 to 7 days after insemination for the quantification of perivitelline layer sperm holes. Eggs traits (weight, specific gravity, yolk and dry shell weight) were measured at 30 and 60 days from the start of dietary treatments.

The sperm penetration assay of Bramwell et al. (1995) was used to quantify the perivitelline layer sperm holes. An approximately 1 cm2 section of the perivitelline layer above the germinal disc was cut free, cleaned, mounted to a microscope slide, fixed, and stained with

Schiff's Acid reagent to generate a contrast with the sperm holes. The holes were counted at 100X magnification. The raw numbers and change in numbers over time were used as a representation of quantifiable fertilization potential.

Observations

Egg Traits: The use of Sel-Plex® rather than sodium selenite as the dietary selenium source has previously been shown to increase shell breaking strength after 42 days in laying hens at 80 wk of age (Paton et al., 2000a). Our study found no significant differences in egg traits between the various treatments after 30 days on the diets. However, the Sel-Plex® treatment resulted in numerically the greatest positive change in shell quality during the 30 day period of this trial.

Research with younger birds (26 wk of age) has indicated no difference in shell quality with the use of Sel-Plex® although the comparisons were made after only 28 days on the diet (Paton et al., 2000a).

After 9 wks on the diets, shell weights were higher in the Sel-Plex® group than in the controls, while shell weights in the group given inorganic Se was intermediate. Egg specific gravity, a measure of shell quality, was greater in the Sel-Plex® treatment than in either the control or selenite groups.

While higher dietary selenium levels preserved shell quality to some degree, the organic selenium in Sel- Plex® proved to have a more substantial effect on the preservation of shell quality characteristics.

Perivitelline Sperm Hole Assay

The Sel-Plex® treatment group had the highest mean number of sperm holes in the 2 to 4 days after insemination study period, while controls had the lowest number. Although neither the Sel- Plex® nor the control treatment were statistically different from the inorganic Se treatment, they were statistically different from each other.

A similar relationship among the dietary treatments occurred for the 5 to 7 day period, and for a comparison over the entire 2 to 7 day period. The perivitelline sperm hole numbers declined at a similar rate among dietary treatment groups between the 2 to 4 day and the 5 to 7 day study periods. As the control group started with a lower number of sperm holes than either of the other two treatment groups, their final numbers at 7 days were very low.

The Sel-Plex® and inorganic Se treatments both retained higher fertility potential throughout the sampling period allowing for a longer period of time between artificial inseminations. While not statistically different, the average number of sperm holes at the site of fertilization in the Sel- Plex® group was higher than that of the inorganic Se treatment. This indicates that the greater bioavailability of selenium in Sel-Plex® compared to inorganic Se may be advantageous for the fertility of the female based on changes to the oviduct environment.

Selenium seems to play an important role in the maintenance of fertility in older laying hens. This is most likely due to the seleniumdependent GSH-Px improving the environment of the SST (Surai, 2000). The SST need to maintain a stable environment and the elimination or reduction of free radicals within the tubules is essential.

Summary: Study 1

Selenium supplementation is beneficial to increasing and maintaining fertility and shell quality in older hens. Supplementation with the organic selenium in Sel-Plex® may have a greater impact on reproductive ability than inorganic sodium selenite. Factors such as age and length of exposure to the diet also play a role in the results of both this and past studies.

The form and quantity of dietary selenium appear to impact the oviduct environment of the hen. Fewer sperm are able to survive under low dietary selenium conditions (control) compared to conditions provided by supplementation with an organic selenium source (Sel-Plex®). Conditions in the SST may be central to the differences noted in the number of sperm being able to reach the site of fertilization. Through a combination of a more stable, antioxidant-free environment with a potentially slowed sperm metabolism, more sperm may be able to survive storage. Selenium source appears to influence the hen's contribution to the fertility of the breeder flock.

Effects of dietary selenium source on the fertility and hatchability of broiler breeders.

Supplying selenium to broiler breeders in the organic selenoamino acid form may have an important impact on poultry reproduction at the level of sperm formation, sperm storage, and in the hatching egg through increased protection from oxidative damage.

This experiment was designed to provide information on the role of dietary selenium form on both female and male fertility. While there is evidence on a flock basis that selenium source affects broiler breeder female fertility, it is not as clear how these benefits are being expressed.

Previous work suggests the use of an organic selenium source (Sel-Plex®) can lead to improved egg production, shell quality, sperm viability, and embryo survival. Egg shell quality may be enhanced through an altered efficiency of calcium metabolism, and sperm quality may be enhanced through protective antioxidant effects in the male and in the female oviduct. This study assessed some of these production and fertility traits in broiler breeders maintained individually.

The objective of this trial was to characterise specific effects of dietary selenium source on fertility and embryo viability aspects in commercial broiler breeder stocks. A female diet with no added selenium was used to identify the impact of dietary selenium.

Inorganic and organic dietary selenium sources were compared to demonstrate the impact of differences in selenium accessibility and tissue storage on reproductive traits and embryo survival. Higher rates of production, fertility, and ultimately chick quality, would decrease the number of birds required to maintain current rates of production, as well as the overall cost of production.

Methods

Ross 508 pullets were reared in a light tight facility following the breeder BW profile (Aviagen Inc).

From photostimulation (22 wks of age) pullets were fed a selenium-free laying ration (No added Se), a standard ration containing sodium selenite (0.3 mg Se/kg), or a ration containing selenium yeast (0.3 mg Se/kg from Sel-Plex®).

Thirty hens per treatment were inseminated weekly (from 30 wks) using pooled semen from males fed a standard, sodium selenite diet or a diet containing the same amount of Se from Sel-Plex®. Individual egg production to 58 wk, egg weight, egg specific gravity, and BW were recorded. At 35 and 57 wk of age, eggs from 2 to 5 days after insemination were subjected to the perivitelline sperm penetration assay to measure the number of sperm penetrations near the germinal disk. Eggs were incubated weekly and the hatch residue broken out to determine fertility, hatchability, and embryonic mortality.

Observations

Sperm Management: Perivitelline sperm hole numbers of Sel-Plex® and selenite treatment eggs were similar. Both treatments had more sperm holes than eggs from unsupplemented hens by a factor of 2 to 3. Sel-Plex® supplementation improved maintenance of sperm numbers between the day 2 and the day 5 sampling.

By day 5, Sel-Plex® eggs still had an average of 60 perivitelline sperm holes compared to 14 in control eggs, while selenite treatment eggs were intermediate (31 holes). These values represented a decline of 31% in apparent viable sperm population in Sel-Plex® birds between Day 2 and 5 after insemination compared to a 46% and 48% drop within non-supplemented and selenite-fed birds, respectively.

The ability to maintain a viable sperm population for as long as possible reduces necessary frequency of insemination. While selenium appears essential to allow the sperm into the oviduct, organic selenium in Sel-Plex® may have an advantage over inorganic selenium in keeping the sperm population stable and alive. This is especially important as the hens age and have a reduced sperm storage capacity at the uterovaginal junction.

The males on the Sel-Plex® diet produced greater semen volume early in production, with an average of 0.36 ml/bird compared to 0.19 ml/bird in males on the selenite diet (36 weeks of age). At 56 weeks of age, this difference was no longer significant, but remained at nearly the same magnitude. The comparison was complicated at the later ages due to several small males dropping out of semen production part way through the trial (selenite treatment). Testes of all birds are currently being examined for the presence of functionally active sperm producing cells.

Egg Production and Egg Quality Traits

Birds on the Sel-Plex® diet entered egg production slightly behind the other feeding treatments (nonsignificant difference), but caught up within a few weeks. Early egg production to 29 wk of age was not different. In fact, the rate of lay was similar through most of the production period.

However, during the late lay period (49-58 weeks) the hen-housed rate of lay was 68% in Sel-Plex® birds compared to 61% and 60% in the selenite and non-supplemented treatments, respectively. The Sel-

Plex® birds produced an extra 5 eggs/bird during this period, on average. This is an important time to be producing more eggs, as egg size is higher than in young breeders, which results in a larger chick size and ultimately a greater broiler weight. Edens (2002) also indicated that the egg production of Sel-Plex®- fed hens initially lagged behind, but caught up and even surpassed that of the selenite-fed hens after 5 wk.

Ultimately the settable egg production in the dietary groups was 168.5 (non-supplemented), 168.6 (selenite), and 174.6 (Sel-Plex®) eggs/bird.

Overall, unsettable egg production ranged from 3.49% in non-supplemented hens to 1.9% in Sel- Plex® hens and was not significantly different. However, during the late lay period (49-58 wk), the Sel-Plex® hens produced significantly fewer unsettable eggs (0.9%) than non-supplemented hens (3.3%), while selenite hens were intermediate (1.7%).

Egg weight and shell quality of settable eggs was assessed throughout the trial and was unaffected. This means that if the hen laid a good egg, it also had a good shell. However, diet affected how many eggs were produced with good shells, as shell defects were the primary egg quality problem in unsettable eggs.

Feeding Sel-Plex® organic selenium to laying hens at 80 wk of age has previously been shown to improve shell breaking strength (Paton et al., 2000a).

Interestingly, dietary selenium affected the change in shell weight as the hens aged. Between 36 and 56 wk of age, shell weight increased by 0.55, 0.80, and 0.76 g in eggs of the non-supplemented, selenite, and Sel-Plex®-fed hens, respectively.

During this time egg size also increased, meaning that the shell was being stretched over more egg, and therefore making up a smaller percentage of total egg weight. The proportion of shell weight dropped by 0.84% of egg weight in non-supplemented hens, 0.80% in selenitefed hens, and 0.57% in Sel-Plex® hens between 36 and 56 wk of age.

The Sel-Plex® hens were significantly less affected by age-related declines in the proportion of egg shell than the non-supplemented hens. While egg specific gravity was not significantly affected, this may be an indicator of increased shell thickness in the Sel-Plex® treatment (not tested). If this were different, there could be implications for incubation success and for defence from contamination in the barn.

Hen body weight followed a similar pattern throughout the production period. However, the non-supplemented hens grew heavier than the other treatment hens by 42 weeks of age. This difference carried through to 58 weeks of age.

This comparison is somewhat artificial, as the body weight profile of the non-supplemented group was inflated by hens that dropped out of lay at a fairly young age. Nutrients they were no longer allocating to egg production went into growth instead. By the end of the trial, 100% of the Sel-Plex® hens were still in active production while only 87% and 90% of the non-supplemented and selenite treatment hens remained, respectively.

Lack of production was due to either birds ceasing production, or to hen mortality (mortality limited to non-supplemented treatment). Reduced mortality has been linked to selenium supplementation (Arnold et al., 1974), particularly under stress conditions such as an immune challenge (Edens, 2001). However, this does not explain the increased proportion of birds still in active lay at 58 weeks of age.

Reproduction in the broiler breeder can be a fragile state and is often the first thing to go when there is stress, or nutrients are insufficient. The fact that all Sel-Plex® birds were still in production may relate to an improved efficiency of nutrient uptake. Presumably the gut benefits from the improved protection from cell membrane damage afforded by the organic selenium. During the late production period, these birds were producing more eggs than hens of the other treatments with no effect on their body weight relative to that of hens on the other treatments.

Fertility, Hatchability, and Embryonic Mortality

Prior to 34 weeks, hatchability averaged 88% in Sel- Plex® treatment eggs compared to 80% in selenite-fed birds and 77% in non-supplemented birds, and was similar in all treatments after peak production.

Overall, fertility, hatchability, and hatch-of-fertile eggs demonstrated the beneficial nature of dietary selenium, but did not differentiate between selenium sources. Fertility, for example, was 86.9% in non-supplemented hens compared to 90.1% in hens on selenite and Sel-Plex® diets. Not including selenium in the diet did not seriously harm hatchability, which is in contrast with work by Latshaw and

Osman (1974) demonstrating a drop in hatchability to 18% in seleniumdeficient hens. The current study may have provided more naturally occurring selenium in the other feed ingredients and the non-supplemented dietary treatment was not imposed until photostimulation (22 weeks of age).

Embryonic mortality can be a telling identifier of specific dietary or genetic effects. Problems with early embryonic mortality (1-14 days of incubation) can point to nutrient deficiencies. In this study, 5.33% of non-supplemented embryos died during this period compared to 3.72% (selenite) and 3.52% (Sel-Plex®) in the selenium-supplemented hens. While selenium source did not make a difference here, clearly selenium supplementation was shown to be important.

A beneficial effect of organic selenium was expected for the late incubation and hatch period, as this is the time of the greatest oxidative load for the embryo, and when the protective antioxidant effects of the Sel-Plex® may be most apparent.

Variability among birds reduced the significance of this comparison, however, and late embryonic mortality, dead-in-shells, and hatchery culls totaled 3.66%, 3.85%, and 3.14% of eggs set for the nonsupplemented, selenite, and Sel-Plex® treatments, respectively. Based on these numerical differences, there appears to be a potentially protective effect of Sel-Plex® compared to inorganic selenium in the diet.

Examining this relationship more closely revealed an interesting trend over time. Late embryonic mortality of all treatments was similar at the start of the trial, when all birds were still on a fairly high plane of nutrition. As the hens aged, late embryonic mortality stayed almost constant in non-supplemented and selenite hens, while it decreased in Sel-Plex® hens. As feed allocations were reduced with age, the micronutrients would have been in shorter supply.

The improved efficiency of selenium uptake in the Sel-Plex® diet may not have made a substantial difference on hatchability until a nutrient challenge was faced by the flock. This fits with observations that Sel-Plex® can demonstrate benefits in stressful situations. Heat stress and long-term egg storage are examples of stress situations where Sel-Plex® has been shown to help. Surai and Dvorska (2001) indicate that there are numerous on-farm stress conditions that could be alleviated in part by organic selenium supplementation.

Ultimately what determines the success of a broiler breeder management programme is chick production. In this trial, chick production was calculated from the hatchability of settable eggs. The unsupplemented hens produced an average of 131.3 chicks/hen-housed by 58 weeks of age, while selenite hens produced 139.1 chicks/hen, and Sel-Plex® hens produced 145.3 chicks/hen. Between the selenite and Sel- Plex® selenium source diets, the numerical differences in settable eggs, embryonic mortality, hatchability, and hatch of fertile culminated in a difference of 5.8 chicks in favour of the Sel-Plex® hens.

Reproductive traits were improved with the inclusion of dietary selenium, while Sel-Plex® supplementation also improved sperm survival in the oviduct, as well as settable egg production late in lay through increased egg production and reduced shell defects. Ultimately, chick production was improved in the Sel-Plex® treatment through more successful settable egg production and the additive culmination of numerical improvements in hatchability and embryo viability measurements. Selenium is essential in the diet for a successful reproductive effort. Additional benefits of using the Sel-Plex® are also possible. Selenium source appears to influence the hen's contribution to the fertility of the breeder flock and to beneficially affect semen volume early in production.

Managing the broiler breeder female for optimal chick production requires an understanding of reproductive physiology, nutrition, and their interaction. Besides a thorough knowledge of everyday management, there must also be an awareness of feed ingredients and their interactions both with each other and with environmental effects.

Whereas the basic composition of the egg is fairly constant, diet and specific feed ingredients can affect what and how much of some of the minor ingredients make it into the egg and ultimately the embryo. Specialised feed ingredients are available that behave differently than traditional ingredients and can enhance egg and chick quality under the right conditions. Together these factors can be used to enhance embryo survival.

Reproduction

The avian reproductive system is heterosexual and requires both a male and a female, each to contribute half of the genetic constitution of the offspring. The male contributes his half by way of the sperm produced by the testes and carried in the semen. The female contributes hers in the ovum carried by the egg yolk produced by the ovary. The ovum is often referred to as the blastodisc, blastoderm or germ disc. After release from the follicle on the ovary, the yolk moves into the oviduct where it is fertilized and has added to it the albumen, shell membranes and shell.

Male Reproductive System

The male reproductive organs in the domestic fowl consist of two testes, each with a deferent duct that leads from the testes to the cloaca. Fowls do not have a penis found in other animals. The testes are bean shaped bodies located against the backbone at the front of the kidney. Their size is not constant and they become larger when the birds are actively mating. The left testes are often larger than the right. On the inside of each as they are located in the body cavity is a small, flattened area that is believed to correspond to the epididymis of mammals. The deferent duct starts at this flattened area.

Deferent Duct

The deferent duct transports the sperm from the testes where they are formed to the cloaca from which they enter the oviduct of the female when mating. The deferent duct enters a small pimple like structure in the cloaca. This structure represents the mammalian penis and is much larger in ducks to form a penis like organ. The deferent duct is quite narrow at first but widens as it approaches the cloaca.

Testes and Sperm

In the testes very twisted tubes called seminiferous tubules are found. It is in these tubules that a special process of cell division called meiosis and transformation produces the sperm. The sperm carry half of the total chromosomes required to produce an embryo. The mother provides half the other. One cubic millimetre of the fluid called semen produced by the male contains on average 3-5 million sperm. Under a microscope the sperm of the fowl will be seen to have a long pointed head with a long tail. The testes also produce hormones called androgens that influence the development of what are called secondary sex characteristics such as comb growth and condition, male behaviour and mating.

Female Reproductive System

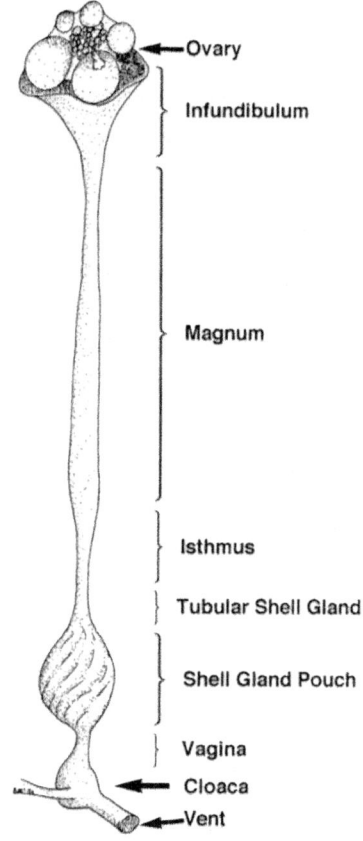

Figure 1: Diagram of the Oviduct

The female reproductive system in the domestic fowl consists of the ovary and the accompanying oviduct. While the female embryo chicken has two sets of reproductive organs, only one of these, the left survives and reaches maturity to produce eggs. The single surviving ovary is located in the laying hen just in front of the kidneys in the abdominal cavity and is firmly attached to the wall of the cavity. The ovary is well endowed with blood vessels to ensure there is no hindrance to the transport of nutrients to the developing yolk.

Ovary

The ovary consists of a mass of yellowish, rounded objects called follicles, each containing an ovum or yolk. There are many such follicles but only a small number in comparison will ever reach maturity to produce an egg. When the hen is in lay the ovary will be active. The size of the follicles will vary from very small to those approaching the normal yolk size in the egg – up to 40 millimetres in diameter and containing a fully matured yolk ready for release into the oviduct.

It is possible to find five stages of development in the active ovary:

1. Primary follicles – follicles that have not yet commenced to grow
2. Growing follicles
3. Mature follicles – follicles ready or nearly so for release
4. Discharged follicles – where the yolk has just been released
5. Atretic follicles – those from which the yolk has been released some time ago.

Yolk

It takes approximately 10 days for a yolk to develop from the very small to the normal size found in eggs and during this time it is contained in the follicle. The follicle acts as a sack during this period of development supplying it with the nutrients required for its growth. When a mature follicle is examined an elongated area virtually free of blood vessels will be found on the distal surface of it. This area, called the stigma, is where the follicle normally splits to release the yolk into the oviduct. If, for some reason, the follicle splits at other than the stigma, the numerous blood vessels that rupture will result in free blood being found in the egg i.e. a blood spot will form.

Oviduct

The function of the oviduct is to produce the albumen, shell membranes and the shell around the yolk to complete the egg. It is a long tube well supplied with blood via numerous blood vessels. There are many glands found in its walls that produce the albumen, the shell membranes and the shell. In the non-layer the oviduct is quite short and small in diameter. However, once the reproductive system becomes active, it grows to a length of 70-80 centimetres with a variable diameter depending on the function of the section being examined.

The oviduct consists of five distinct parts or sections, each having different functions:

1. Infundibulum (or funnel) – located adjacent to the ovary and with long segments enclosing the ovary, the infundibulum collects the yolk after it's release from the follicle as a funnel

and directs it into the oviduct. This section has very thin walls and is 6-9 centimetres long. Fertilization of the ovum by the male sperm occurs here.

2. Ampulla or magnum – at approximately 40 centimetres long it secretes more than 40% of the albumen.

3. Isthmus – at about 12 centimetres in length, it secretes some albumen and the shell membranes.

4. Uterus or shell gland – at approximately 12 centimetres in length it secretes about 40% of the albumen and the egg's shell.

5. Vagina – at approximately 12 centimetres in length, it secretes the egg's outer cuticle and possibly the shell pigment.

Androgen, Oestrogen and Progesterone

In addition to the production of eggs, the female reproductive system also produces hormones that aid in the control of body functions. These include androgen, oestrogen and progesterone. Androgen causes comb growth and condition, and has a function in the formation of albumen. Oestrogen causes the growth of the female plumage, mating and nesting behaviour, oviduct development together with the nutrient supply to the ovary/oviduct for egg formation. Progesterone, with androgen is involved in the production of albumen and the carriage of the message to the pituitary gland to release luteinising hormone. The female reproductive system remains dormant in the young chicken and growing pullet until she reaches the age when these organs start to prepare for the normal production of eggs. One of the first signs of her developing maturity is the change in the comb development. This organ starts to grow and to take on a vivid red hue as the hormones produced by the now awakening ovary start to have an effect.

The Formation of the Hen's Egg

The normal egg consists of the following major parts:

1. Yolk carrying the ovum – produced by the ovary

2. Albumen or white – produced mainly in the magnum

3. Shell membranes – produced in the isthmus

4. Shell – produced in the uterus or shell gland.

The Ovary and Yolk Formation

The ovary is attached to the abdominal cavity wall by the meso-ovarian ligament. It carries anything from 2000 to 12000 small ova in miniature follicles on its surface, plus hormone producing cells in its body. Not all of the ova found on the immature ovary develop and only approximately 200 to 350 reach maturity under normal modern commercial practice. Each yolk or ova takes about 10 days to grow and reach maturity when it is approximately 31% of the weight of the egg.

Table 1: The composition of the yolk material

Component	%
Water	48.0
Protein	17.5
Fat	32.5
Carbohydrate	1.0
Other compounds	1.0

The yolk is laid down in concentric rings of darker and lighter coloured material, the colour being produced by xanthophylls that are yellow/orange/red pigments occurring in many plants, plant products and other naturally occurring materials. The bulk of the yolk material provides a source of food for the developing embryo that originates by the fertilizing of the germ disc or blastoderm usually located on the upper surface of the yolk of the broken out egg. It lies in the surface segment of the latebra which is a vase shaped segment of different yolk with it's base in the centre of the yolk, the lips on the surface and the stem joining the base to the lips.

The yolk is contained in a very thin, transparent membrane called the vitelline membrane. As an egg becomes stale, the vitelline membrane becomes significantly weakened and often breaks to release the yolk contents when the stale egg is broken out. On ovulation the yolk is released and enters the oviduct where, as it passes along that organ, fertilization occurs and the remaining parts of the egg are added around it. The yolk is located on the ovary held in a sack called the follicle. The follicle, while being quite thin walled, is extremely well supplied with blood vessels. These are necessary to carry the materials constituting the yolk that have been formed in the liver.

Yolk development in the maturing pullet is initiated by follicle stimulating hormone (FSH) produced by the anterior lobe of the pituitary gland. The compounds in the yolk material are formed in the liver and, on the appropriate signal, are transported by the blood stream to the target follicle and into the yolk. The appropriate signal for this development comes from hormones – oestrogen, progesterone and testosterone produced by the ovary after receiving the signal of the Follicle Stimulating Hormone. These ovarian hormones also provide the stimulus for the formation of the development of the oviduct.

Ovulation

The release of the yolk – the process of ovulation, is the major controlling factor influencing the subsequent steps in the formation and laying of the egg. As a consequence, factors influencing ovulation are of critical importance to the various aspects associated with egg production. The presence of a mature yolk in a follicle causes hormones from the ovary to stimulate the release of luteinising hormone (LH) by the pituitary gland. The presence of LH in the blood stream causes the follicle containing the mature yolk to split along the stigma thus releasing it into the oviduct abdominal cavity adjacent to the oviduct.

Sex Maturity

Sex maturity is reached when the hen lays the first egg in her life. While generally sex maturity is genetically controlled, environmental factors play a very significant role. It will be in the age range of 18-24 weeks depending on genotype, but it can be manipulated by controlled feeding practices, light intensity and day length management and other management practices.

Initiation of Ovulation

The controlling mechanism setting the time of the day for the first ovulation is not fully understood. However, nervous and hormonal factors are important. Subsequent ovulations are, however, controlled largely by the time of the previous egg passing through the vent i.e. being laid. Subsequent yolk release, if at all, occurs approximately 40-60 minutes after the previous egg has been laid.

Clutches

Eggs laid on successive days are called a clutch. Clutches are separated by days when no eggs are laid. Clutch size is an individual characteristic and may vary in a flock from 2 up to 100 eggs. However, the normal clutch size is significantly less than that and ranges from 3-8 eggs. The larger the clutch size the better will be the total production. Small clutch size indicates an inferior layer and is usually associated with long breaks between them.

Egg Formation Time

The time taken from ovulation till when the egg passes through the vent varies with individuals within the range 23-26 hours. If the time is longer than 24 hours then the time of laying will be progressively later in the day for each successive egg in the clutch. When eggs are laid at a late hour, an ovulation is missed and the start of a new clutch will be earlier in the next laying day.

Ovulation Time

Hens that produce long clutches release the yolk very shortly after first light (whether natural or artificial light). Successive ovulations occur very shortly after the laying of the previous egg. Those producing short clutches usually release the yolk later in the day and often have longer periods between laying time and the next ovulation.

Laying Pattern

When pullets first commence to lay their hormonal and other controlling systems have not yet reached a state of balance. As a consequence, the first eggs are laid in a somewhat haphazard sequence. However, once these systems have reached a state of balance (usually after 7-10 days), egg production becomes more regular. Peak ovulation is reached 3-5 weeks after first egg. This will be held for a period and then will decline steadily thereafter until the bird moults or some other factor causes a cessation of production for a period.

Oviduct

The remainder of the egg i.e. the albumen, the shell membranes and the shell are produced by different segments of the oviduct. These segments are:

- Infundibulum
- Magnum or ampulla
- Isthmus
- Uterus or shell gland
- Vagina
- Cloaca.

In the egg laying hen the oviduct is a tube like organ consisting of the previously named segments with one end lying adjacent to the ovary and the other entering the vent. It is approximately 70 centimetres long and is very glandular. The glands of the different segments produce the remaining different parts of the egg. Because of its function the oviduct is very well supplied with blood vessels.

Infundibulum

This segment is funnel shaped and lies adjacent to the ovary. It is up to 9 centimetres long in the laying hen and has the function of searching for and engulfing the yolk that has just been released from the follicle into the adjacent ovarian pocket or body cavity. The yolk remains in the infundibulum for about 15 minutes and it is in the infundibulum that fertilization takes place.

If the infundibulum should malfunction and not engulf the yolk, the yolk will remain in the ovarian pocket from where normally they will be absorbed within three days. If the number of such occurrences reaches a high level, the yolks will accumulate in the ovarian pocket faster than they can be absorbed. Such birds' are called internal layers – the abdomen becomes distended and the hens adopt a very upright stance.

Magnum or Ampulla

The magnum is the longest segment at up to 40 centimetres long. Its function is to add approximately 40% of the albumen to the developing egg that takes about three hours to move through. These percentages vary considerably depending on factors including the genetics of the hen, age of the bird, the egg's age and/or storage conditions. However, in a good quality, freshly laid egg the above relationship mostly applies.

Table 2: Different types of albumen in the normal egg

Albumen type	%
Chalazae and the chalaziferous layer	2.7
Liquid inner layer	17.3
Dense layer	57.0
Outer liquid layer	23.0

The chalazae are two twisted chords of albumen extending from the opposite sides of the yolk into the remaining albumen in the broken out egg. These two cords extend into the ends of the egg along the longitudinal axis and are parts of a very thin envelope of special albumen that surrounds the yolk and holds it in its position. The yolk has to remain centrally located for the survival of the embryo. The yolk turning or rotating as it passes along the oviduct causes the twisted effect of the chalazae.

While the bird produces only dense albumen, as the egg moves along the oviduct, water is added thus making liquid albumen. The rotation of the developing egg causes the albumen to separate into different layers – the inner liquid and the dense layers. The outer liquid layer is caused by the addition of more water when in the uterus. The dense layer contains significant amounts of mucin that binds it together in a jelly like form.

- As an egg stales, the amount of dense albumen decreases as it changes to the liquid form. The liquid form increases in volume and becomes even more fluid.

Isthmus

The isthmus is approximately 12 centimetres long and has the functions of adding approximately 20% of the albumen and the shell membranes to the egg. There are two shell membranes:

1. The inner shell membrane – laid down first
2. The outer shell membrane – laid down last and about three times the thickness of the inner membrane.

The isthmus takes approximately 75 minutes to carry out its tasks. While the egg is still in the oviduct the shell membranes appear as one over the total surface of the egg so close are they associated with each other. However, as the egg cools after it has been laid, the membranes separate, usually at the larger end to form the air cell. The air cell in the new laid egg is approximately 1.5 centimetres in diameter and approximately 0.5 centimetres deep.

As the egg ages, the interior contents lose water and the air cell increases in size. This change in size is an indicator of egg quality as related to the age of the egg and the holding conditions. The shell membranes consist of a fibrous protein material and act as a barrier to the penetration into the egg by bacteria and fungi. They also help reduce the rate of evaporation of water from the egg thus slowing the rate of deterioration of the egg.

- The isthmus also lays down the foundation for the shell by forming the first crystals of calcium carbonate on the outer shell membrane.

Uterus (Shell Gland) and Eggshell Quality

The uterus is a relatively short, bulbous gland up to 12 centimetres in length. The developing egg remains in the uterus for 18-20 hours while approximately 40% of the albumen and all of the shells are added. It is for this reason that the organ is often called the shell gland. Shell formation really begins by the deposition of small clusters of calcium carbonate crystals onto the outer shell membrane while in the isthmus. These are the initiation grains for the subsequent calcium carbonate deposition in the uterus. The number of these grains is genetically controlled and is related to the subsequent shell thickness – the more grains deposited in the isthmus, the thicker will be the final shell.

The shell of an egg is formed in two layers:

1. Mammillary layer: a sponge like layer composed of soft calcite crystals ($CaCO_3$). This layer is the inner layer.

2. Palisade layer: formed of columns of hard calcite crystals – the longer the columns the stronger the shell. This layer is the outer layer.

The calcium for the eggshell comes from three sources – the diet, special bone called medullary bone (found in the cavity of long bones) of point of lay pullets and the skeleton. The hen uses approximately 2.5 grams of calcium in the formation of one normal egg. She cannot absorb sufficient calcium from her diet each day (approximately 2.0 grams per day) to supply this need and hence, it becomes necessary for her to utilise skeletal calcium to make up the shortfall. This is particularly so at night when most of the shell is being formed but the hen in unlikely to be eating. In addition to the calcite, the shell also contains small quantities of sodium, potassium and magnesium.

The carbonate ions go with the calcium to form the calcium carbonate of the egg's shell come from the blood and from the shell gland. If anything should interrupt the supply of carbonate, thin-shelled eggs will result. This occurs in hot weather when hens pant to remove excess heat energy. The increased respiratory rate removes carbon dioxide from the blood thus reducing the carbonate ions available for eggshell formation.

There are many factors that influence eggshell quality:

1. *Length of Time in Lay:* The longer the bird is in lay, the weaker the shells will become because of her inability to obtain daily

enough calcium from her diet to supply all of her needs for one egg. As a consequence, the better layers will deplete their skeleton.

2. *Increased Environmental Temperature:* This results in reduced food consumption (and calcium) and the reduction of carbonate ions because of panting.

3. *Egg Laying Time:* Eggs laid early in the morning are more likely to have thinner shells than those laid by the same bird later in the day. This is because in the case of those eggs laid early the shells have been deposited during the hours of darkness when the bird doesn't eat i.e. consume calcium for the shell.

4. *Stress:* Stressed birds lay thinner shelled eggs.

5. *Body Checked and Misshapen Eggs:* Most of these defects are caused by the birds being startled shortly after the egg has entered the uterus and the first layers of calcium carbonate have been deposited. At this stage the shell is very fragile and weak and when startled the hen's muscles contract (including those in the wall of the uterus) and thus crack the newly forming shell. These are covered by subsequent depositions of shell but the damage remains in the form of body checks and/ or misshapen eggs.

6. *Disease:* Certain diseases, e.g. infectious bronchitis cause weak shell and misshapen eggs.

7. *Drugs:* Certain drugs influence eggshell formation and deposition.

The shell of an egg contains openings or pores. There are approximately 8000 such pores in the shell of a normal hen's egg. The function of these pores is to provide for the gaseous exchange during incubation and embryonic development.

The developing embryo requires oxygen and gives off carbon dioxide. When the egg is first laid most of the pores are closed. However, as the egg ages more and more pores open up. The cuticle deposited on the outer shell is composed of organic material and water and blocks the pores. During the laying process the cuticle acts as a lubricant, but once laid, the egg's surface soon dries and the residue – mainly protein closes off most of the pores as a barrier to the invasion of bacteria and fungi.

Vagina

The vagina is about 12 centimetres in length. While not known for sure, it may have the function of adding pigment to the outer shell to provide the egg with its colour.

Cloaca

The egg is held in the cloaca immediately prior to being laid. It may be in the cloaca for several hours, but usually is held there for a much shorter time. Although the egg usually enters this organ small end first, it usually rotates there to be laid large end first. However, if the bird should be startled at this time the egg may be forcibly expelled small end first.

Summary of the Reproductive System

The domestic fowl's reproductive system, while functioning in some ways in a manner similar to that of mammals, is very different in many aspects. Evolution has given the birds a reproductive system where the offspring are separated from the mother to enhance her ability to fly and reproduce at the same time. The formation of an egg is a very complex activity during which much can go wrong. The quality of the final product, the egg as it is laid is influenced by both genetic and management factors. The efficient manager, to be able to produce eggs efficiently whether for incubation or for table use, should have a good knowledge of the reproductive system and the formation of the egg.

Ovulation

Ovulation is the process in a female's menstrual cycle by which a mature ovarian follicle ruptures and discharges an ovum (also known as an oocyte, female gamete, or casually, an egg). Ovulation also occurs in the estrous cycle of other female mammals, which differs in many fundamental ways from the menstrual cycle. The time immediately surrounding ovulation is referred to as the ovulatory phase or the periovulatory period.

Overview

The process of ovulation is controlled by the hypothalamus of the brain and through the release of hormones secreted in the anterior lobe of the pituitary gland, luteinising hormone (LH) and follicle-stimulating hormone (FSH). In the pre-ovulatory phase of the menstrual cycle, the ovarian follicle will undergo a series of transformations called cumulus expansion, which is stimulated by FSH. After this is done, a hole called the stigma will form in the follicle, and the ovum will leave the follicle through this hole. Ovulation is triggered by a spike in the amount of FSH and LH released from the pituitary gland. During the luteal (post-ovulatory) phase, the ovum will travel through the fallopian tubes towards the uterus. If fertilized by a sperm, it may perform implantation there 6–12 days later.

Ovulation occurs when a mature egg is released from the ovary into the abdominal cavity. Afterwards, it will eventually become available to be fertilized. Concomitantly, the lining of the uterus is thickened to be able to receive a fertilized egg. If no conception occurs, the uterine lining as well as blood will be shed in menstruation.

In humans, the few days near ovulation constitute the fertile phase. The time from the beginning of the last menstrual period (LMP) until ovulation is, on average, 14.6 days, but with substantial variation both between women and between cycles in any single woman, with an overall 95% prediction interval of 8.2 to 20.5 days.

Cycle length alone is not a reliable indicator of the day of ovulation. While in general an earlier ovulation will result in a shorter menstrual cycle, and vice versa, the luteal (post-ovulatory) phase of the menstrual cycle may vary by up to a week between women.

Details

Strictly defined, the ovulatory phase spans the period of hormonal elevation in the menstrual cycle. The process requires a maximum of thirty-six hours to complete, and it is arbitrarily separated into three phases: pre-ovulatory, ovulatory, and postovulatory.

Prerequisite Events

Through a process that takes approximately 375 days, or thirteen menstrual cycles, a large group of undeveloped *primordial follicles* dormant in the ovary is grown and progressively weaned into one *pre-ovulatory follicle*. Histologically, the pre-ovulatory follicle (also called a mature Graafian follicle or mature tertiary follicle) contains an oocyte arrested in prophase of meiosis I surrounded by a layer of corona radiata granulosa cells, a layer of mural granulosa cells, a protective basal lamina, and a network of blood-carrying capillary vessels sandwiched between a layer of theca interna and theca externa cells. A large sac of fluid called the antrum pre-dominates in the follicle. A "bridge" of cumulus oophorus granulosa cells (or simply cumulus cells) connects the corona-ovum complex to the mural granulosa cells.

The granulosa cells engage in bidirectional messaging with the cells and the oocyte to facilitate follicular function.

By the action of luteinising hormone (LH), the pre-ovulatory follicle's theca cells secrete androstenedione that is aromatised by mural granulosa cells into estradiol, a type of estrogen. In contrast to the other phases of the menstrual cycle, estrogen release in late follicular phase has a stimulatory effect on hypothalamus gonadotropin-releasing hormone (GnRH) that in turn stimulates the expression of pituitary LH and follicle stimulating hormone (FSH).

The building concentrations of LH and FSH marks the beginning of the pre-ovulatory phase.

Pre-ovulatory Phase

For ovulation to be successful, the ovum must be supported by both the corona radiata and cumulus oophorous granulosa cells. The latter undergo a period of proliferation and mucification known as cumulus expansion. Mucification is the secretion of a hyaluronic acid-rich cocktail that disperses and gathers the cumulus cell network in a sticky matrix around the ovum. This network stays with the ovum after ovulation and has been shown to be necessary for fertilization.

An increase in cumulus cell number causes a concomitant increase in antrum fluid volume that can swell the follicle to over 20 mm in diameter. It forms a pronounced bulge at the surface of the ovary called the blister.

Ovulatory Phase

Through a signal transduction cascade initiated by LH, proteolytic enzymes are secreted by the follicle that degrade the follicular tissue at the site of the blister, forming a hole called the *stigma*. The cumulus-oocyte complex (COC) leaves the ruptured follicle and moves out into the peritoneal cavity through the stigma, where it is caught by the fimbriae at the end of the fallopian tube (also called the oviduct). After entering the oviduct, the ovum-cumulus complex is pushed along by cilia, beginning its journey towards the uterus.

By this time, the oocyte has completed meiosis I, yielding two cells: the larger secondary oocyte that contains all of the cytoplasmic material and a smaller, inactive first polar body. Meiosis II follows at once but will be arrested in the metaphase and will so remain until fertilization. The spindle apparatus of the second meiotic division appears at the time of ovulation. If no fertilization occurs, the oocyte will degenerate approximately twenty-four hours after ovulation.

The mucous membrane of the uterus, termed the *functionalis*, has reached its maximum size, and so have the endometrial glands, although they are still non-secretory.

Postovulatory Phase

The follicle proper has met the end of its lifespan. Without the ovum, the follicle folds inward on itself, transforming into the corpus luteum (pl. corpora lutea), a steroidogenic cluster of cells that produces estrogen and progesterone. These hormones induce the endometrial glands to begin production of the proliferative endometrium and later into secretory endometrium, the site of embryonic growth if fertilization occurs. The action of progesterone increases basal body temperature by one-quarter to one-half degree Celsius (one-half to one degree Fahrenheit). The corpus luteum continues this paracrine action for the remainder of the menstrual cycle, maintaining the endometrium, before disintegrating into scar tissue during menses.

Ovulation is the release of a single, mature egg from a follicle that developed in the ovary. It usually occurs regularly, around day 14 of a 28-day menstrual cycle. Once released, the egg is capable of being fertilized for 12 to 48 hours before it begins to disintegrate. Although there are several days of the month in which a woman is fertile, she is most fertile during the days around ovulation.

Clinical Presentation

The start of ovulation can be detected by various signs. Because the signs are not readily discernible by people other than the woman herself, humans are said to have a concealed ovulation. In many animal species there are distinctive signals indicating the period when the female is fertile. Several explanations have been proposed to explain concealed ovulation in humans.

Women near ovulation experience changes in the cervix, in mucus produced by the cervix, and in their basal body temperature.

Furthermore, many women also experience secondary fertility signs including Mittelschmerz (pain associated with ovulation) and a heightened sense of smell.

Many women experience heightened sexual desire in the several days immediately before ovulation. One study concluded that women subtly improve their facial attractiveness during ovulation and period.

Symptoms related to the onset of ovulation, the moment of ovulation and the body's process of beginning and ending the menstrual cycle vary in intensity with each individual woman but are fundamentally the same. The charting of such symptoms, primarily Basal Body Temperature, Mittelschmerz and cervical position is referred to as the Sympto-Thermal method of Fertility Awareness, which serves to allow auto diagnosis by a woman of her state of ovulation. Once training has been given by a suitable authority, fertility charts can be completed on a cycle by cycle basis to chart ovulation giving the possibility of using the data to predict fertility for both natural contraception and pregnancy planning.

The moment of ovulation has been photographed.

Follicular Waves

Research spearheaded by Baerwald *et al.* suggests that the menstrual cycle may not regulate follicular growth as strictly as previously thought. In particular, the majority of women during an average twenty-eight day cycle experience two or three "waves" of follicular development, with only the final wave being ovulatory. The remainder of the waves are anovulatory, characterised by the developed pre-ovulatory follicle falling into atresia (a major anovulatory cycle) or no pre-ovulatory follicle being chosen at all (a minor anovulatory cycle).

The phenomenon is similar to the follicular waves seen in cows and horses. In these animals, a large cohort of early tertiary follicles develop consistently during the follicular phase of the menstrual cycle, suggesting that the endocrine system does not regulate folliculogenesis stringently. Researchers of fertility awareness or natural family planning methods discovered follicular waves in the 1950s. These methods of family planning have always taken multiple follicular waves into account, and this research does not challenge their effectiveness.

The gene Lrh1 appears to be essential in regulating ovulation.

Induction and Suppression

Induced Ovulation

Ovulation induction is a promising assisted reproductive technology for patients with conditions such as polycystic ovary syndrome (PCOS) and oligomenorrhea. It is also used in in vitro fertilization to make the follicles mature prior to egg retrieval. Usually, ovarian stimulation is used in conjunction with *ovulation induction* to stimulate the formation of multiple oocytes. Some sources include *ovulation induction* in the definition of *ovarian stimulation.*

A low dose of human chorionic gonadotropin (HCG) may be injected after completed ovarian stimulation. Ovulation will occur between 24–36 hours after the HCG injection.

Suppressed Ovulation

Contraception can be achieved by suppressing the ovulation.

The majority of hormonal contraceptives and conception boosters focus on the ovulatory phase of the menstrual cycle because it is the most important determinant of fertility. Hormone therapy can positively or negatively interfere with ovulation and can give a sense of cycle control to the woman.

Estradiol and progesterone, taken in various forms including combined oral contraceptive pills, mimics the hormonal levels of the menstrual cycle and engage in negative feedback of folliculogenesis and ovulation.

Animals

- In cats, rabbits, and camelids, ovulation is induced mechanically by the male through copulation.
- Chickens have an ovulation almost every day.
- The embryos of some Marsupial species enter embryonic diapause (or *delayed implantation*) after fertilization.
- After ovulation, oocytes remain in Meiosis I. The oocyte proceeds to Meiosis II before fertilization.

Does circadian clock regulate clutch-size in birds? A question of appropriatness of the model animal.

The Questions of Animal Models

There are some very good reasons why much of biology is performed in just a handful of model organisms. Techniques get refined and the knowledge can grow incrementally until we can know quite a lot of nitty-gritty details about a lot of bioloigcal processes. One need not start from Square One with every new experiment with every new species. One should, of course, occasionally test how generalisable such findings are to other organisms, but the value of models is hard to dispute.

Which organisms have been chosen as laboratory models often depends on contingencies of history. Somebody at some point in time had a large supply of a particular species, did some good research on it and everything took off from there. Some species proved untractable and were discarded.

Others flourished and were adopted by more and more researchers over time. Some models are really good for studying particular areas of biology. The fruitfly is unparalleled as the organism for genetic studies. However, its development is quite unusual so the findings are difficult to generalise even to other Dipterans, even less to other insects, arthropods or other animals.

A good animal model should be one about which there is quite a lot of background information. It should be available in large numbers and it should be rather small in size so a lab can keep large numbers of individuals in a relatively small space. Being social also makes housing easier. A good model animal is also easy to keep in captivity, it is easy to feed and easy to handle without too much danger to the technicians. Ideally, it easily breeds in captivity, has a short generation time and large numbers of progeny.

A Brief Aside about Mammalian Models

It is not surprising that early studies in anatomy and physiology utilised domesticated animals. Much of 18th and 19th century research was done in dogs, cats, rabbits, sheep, goats, pigs and even horses. Such animals were easily available, there was a lot of existing knowledge about them, and they were easy to house, feed and handle. Furthermore, one could always "sell" such research as useful for advances in agriculture.

However, there are definite drawbacks in using these species as well. They tend to be large, with long generation times and with few progeny. Also, being human companions for thousands of years, people get easily attached to them and the anti-vivisectionist movement was quite strong in the 19th century, especially in England.

Thus, after several decades of effort, several new species were domesticated in order to provide the researchers with more tractable and less controversial animals to work with. These include rats, mice, hamsters, guinea pigs, ground squirrels and ferrets. Much of the research in the 20th century was performed on these species.

In the last couple of decades, one of those species gained prominence - the mouse - due to its tractability to genetic techniques. Much of the findings from rodent (including mouse) studies have been found to be generalisable to other rodents, to other mammals (including humans) and to other vertebrates. Much has been learned that proved applicable to endangered species and their reproduction in captivity. Rarely is any work done on rodents these days that as its main aim has the better understanding of the rodents. Mainly, the mouse and the rat are the stand-ins for humans and the research is thought of as biomedical research.

Avian Models

The bird models have quite a different history. The readily available species are the domesticated birds, better known as poultry - chicken, quail, duck, turkey, goose, pigeon and the guinea hen. There was never a need to move away from these species, as they are small, social, breed fast and a lot, and are viewed as food (and not as "cute pets") thus they were a less likely target of anti-vivisectionists.

Sure, many biologists studied wild species in the field and in the lab. Even today, much work is done in parrots, European starlings, house sparrows, house finches, crows, etc. But wild species are tough to work with. One has to go out and catch a new batch for each new experiment. Poultry is so much easier to handle and breed in the lab.

The lack of a historical "switch" from one group of species to another had one unfortunate consequence. Research on domesticated species of birds is too easily dismissed as Poultry Science, geared towards increased meat and egg production at farms (what's wrong with that? The findings from such research usually make the life for farm birds better, e.g., less stressful). It is sometimes forgotten that many scientists use poultry species with a different motivation altogether - as a model for wild birds, hopefully generalisable even broader to other vertebrates or even all animals. Many findings from poultry research are now being used to help endangered bird species survive and breed in the wild and in captivity.

Because my lab animal is a domesticated bird, poultry if you wish - the Japanese quail. I will not be unhappy if my research gets picked up by poultry scientists and applied to make the life of farm birds easier, but that is not my main motivation. I am interested in particular aspects of basic science of chronobiology and I found quail to be an excellent model for questions that are difficult or even impossible to do in mammals. I hope that my work is generalisable to wild birds and will help in saving endangered species from extinction. I hope that at least some of my work will be even more broadly generalisable, perhaps to all vertebrates, or even to all of life.

Japanese quail (*Coturnix japonica*) has been domesticated about 500 years ago in Japan. The name 'Japanese quail' is used to denote the domesticated birds in laboratories and on farms. Their counterpart in the wild, itself not an endangered species, is usually called Asian Migratory Quail, and it appears that there is not that much difference between wild and domesticated populations in many aspects of their biology. With incubation time of 17 days, maturation time of six weeks, and ability to lay almost an egg a day throughout the year, quail is the mouse of the avian biology lab. The chicken genome has been sequenced. Every gene that has beem looked at is between 95 and 100% identical between the chicken and the quail. It appears that the difference between the two species is mainly due to regulatory regions of the genes driving somewhat different trajectoris of embryonic developments, similarly to the difference between humans and chimps.

An important part of my Masters degree research (but not my PhD) was on the mechanisms by which the circadian system controls reproduction - both seasonally (photoperiodism) and daily (the timing of egg-laying). My MS thesis has been published in these two papers.

8

Feeding Broiler Breeders for Chick Quality

For successful broiler production a chick requires good bodyweight, with excellent nutritional reserves at day old. It needs to be in excellent health with a fully functioning immune system. From this starting point, providing the broiler with suitable environment and nutrition will enable optimal performance to be achieved.

The developing embryo and the hatched chick are completely dependent for their growth and development on nutrients deposited in the egg. Consequently the physiological status of the chick at hatching is greatly influenced by the nutrition of the breeder hen.

In reviewing breeder nutrition, it should be remembered that nutrient supply to the broiler breeder is a sum of two parts, namely nutrient content of the diet and quantity of feed supplied to the breeder birds. Both parts need to be balanced to ensure correct daily nutrient supply. It is also very important to realise that the cost of feeding the breeder appropriately to ensure good nutritional status of the chick is very low when viewed on a per chick basis and compared with the total feed cost of raising a broiler to slaughter weight. Calini (2006) calculated that the cost of breeder feed contributing to the production of a chick is equivalent to only 7% of the total feed cost for a broiler grown to 2.5Kg. This illustrates the value of ensuring the best possible nutrition of the breeder.

Nutrient Levels in Broiler Breeder Feeds

When considering nutrient levels in breeder feeds, the nutritionist must focus on the daily supply of individual nutrients to the bird.

Starting with protein, studies have shown that the protein levels fed to breeders in production can affect chick bodyweight and final broiler performance. The relationship between protein content of breeder feed and chick weight seems well defined.

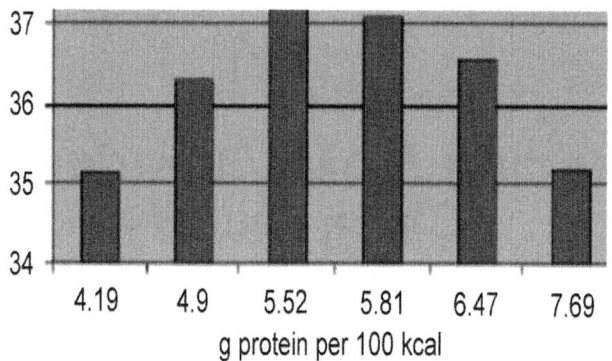

Figure 1: Relationship between Protein Content of Breeder Feed and Chick Weight

Using this information, a breeder diet with an energy density of approximately 2750 Kcal/Kg should have a protein content of 15%. This optimum protein level has been supported by other work, and it is important to remember this is an optimum level, not a minimum, as excess protein can be as detrimental as insufficient protein. In particular, it has been shown that excess protein reduces fertility. Furthermore, consideration must be given to protein quality and the nutritionist must ensure a balance of amino acids is supplied from good quality protein sources.

The impact of energy content of the breeder feed is not as well defined as that of protein. Reviewing studies carried out to evaluate optimum energy intake would suggest that 440 - 480 Kcals/bird/day is most appropriate for optimal chick quality. This equates to 160 - 175 g/bird/day at 2750 Kcal/Kg feed. When considering energy, attention must also be given to fat composition and in particular to the requirement for unsaturated fats such as linoleic acid. This essential fatty acid is required for cell membrane integrity, immune competence and embryonic development, therefore directly affecting chick quality. In practical terms, the inclusion of added fats into breeder feeds should be kept low, with preference for unsaturated fats rather than saturated fats. The major minerals, especially calcium, phosphorous, sodium, potassium, magnesium and chloride are involved in shell formation; improvements in shell quality generally lead to better egg and chick quality. Variations in maternal phosphorous supply have been shown to influence bone ash of young but not older progeny. It is important to supply adequate phosphorus in breeder diets to ensure best possible bone integrity in the early stages of chick growth. In terms of trace minerals, most interest in this field has centred on the use of chelated minerals which have been shown to increase deposition in the egg and transfer to the tissues of the hen and the embryo. Most recent work has focused on the antioxidant status of breeders, embryos, offspring and the role of selenium. Seleno-methionine has been shown to improve both the vitamin E and antioxidative status of eggs, embryos and chicks up to 10 days of age. Supplemental zinc methionine and manganese amino acid complexes have shown improvements in chick immunity and liveability.

Table 1: *Summary of minerals fed to breeders shown to have an effect on progeny performance.*

Fluoride				X
Phosphorous				X
Selenium		X		
Selenomethionine	X		X	
Zinc	X		X	X
Zn-Methionine		X	X	

Vitamins are involved in most metabolic processes and are an integral part of foetal development, therefore the consequence of suboptimal levels of these nutrients in commercial diets are known to result in negative responses to both parent and offspring performance. Vitamins account for about 4% of the cost of a breeder feed, so economising on vitamin inclusion rates is rarely a sensible option. Generally there is a shortage of information on vitamin requirements of broiler breeders especially when related to offspring performance. Most of the breeder work is quite dated and since that time breeder performance has changed.

A review of work on fat soluble vitamins, biotin and pantothenic acid has shown that vitamin E has the largest impact on progeny. In general it seems to be justified to supplement practical breeder feeds with 100 mg/kg vitamin E.

The influence of increased vitamin levels fed to young parent stock on progeny performance is an area which has received significant commercial interest. Internal and field trials have shown that increased vitamin levels (mainly B Vitamins and Vitamin E) improved liveability and early growth. A practical basis for making recommendations is to feed vitamin levels that maximise the resulting level in the egg.

The Influence of Feed Allocation on Chick Quality

Underfeeding the hen can have an impact on chick quality and this is particularly noticeable in the early production period. Modern hybrid parent flocks commence production at a faster rate than in the past and consequently egg output increases over a shorter time span during the early laying period.

Feed allocations during this period have not necessarily increased in line with this egg production trend. Low feed allocation intake by young commercial breeder flocks have been shown to compromise nutrient transfer to the egg, resulting in increased late embryonic death, poorer chick viability and uniformity (Aviagen Ltd 2002). In a study by Leeson (2004) broiler breeders were fed different levels of feed through peak production varying from 140 to 175 grams. Although the increased feed allocation increased bodyweight there was no influence on egg size up to 175g, however chick weight was influenced by feed allocation.

Research shows that nutrient supply to the broiler breeder is of consequence to chick quality and production performance. This places greater emphasis on the nutritionist providing the correct nutrient density diet and the flock manager to provide appropriate feed intake to the bird coming into lay and through the production period.

Broiler Feeding: Mash or Pellet?

Mash or pellet? The decision is not straightforward, because one must take into account numerous factors together with the complex interactions involved in the feed digestion process.

The important factors:
- Genetic evolution and broiler feeding behaviour.
- Feed presentation; feed factory technology; manufacturing costs.
- Housing conditions, management, and health status all have a significant impact on digestion process efficiency and stability,
- Law and regulations with respect to consumer's protection, restriction in the use of antibiotics, growth promoters, and coccidiostats.

Production conditions vary according to the country, and different technical solutions may lead to good techno-economical results:

For instance in Northern Europe (UK; Holland; Denmark), the raw materials are ground very fine to ensure good cohesion throughout the conditioner and expander treatments at 85-90"C. The benefits of this technique are optimised in cases involving very healthy chicks reared on high-level management and biosecurity farms, The technical results are excellent, but the costs are high, in line with the high food safety expectation of the end consumer

Brazil can produce high quality raw materials (maize and soya bean). Because of its geographical position and climate, farms are scattered, often very simple in design and flock density is low. Feed is mainly in coarse mash form (only 30 % of the total feed produced is pelleted).

Between these two examples, there are many ways of achieving the same technical results according to the local technical and economical conditions.

Mash or Pellet? The Question of Genetics

Genetics: Genetic selection has yielded tremendous progress in weight gain. In the 70's-80's when a higher energy feed, in a pelleted form, was made available it helped to better express the real genetic potential of broilers.

Then, as a consequence of faster growth, problems arose that the primary breeders must now take into account in their selection programmes: skeleton disorders (e.g. lameness) and metabolic disorders (e.g. ascites, sudden death). Lately, the tightening of restrictions regarding the use of antibiotics and coccidiostats has given rise to increased incidence of necrotic enteritis.

Practically, the above problems have mainly been addressed through the implementation of lighting programmes. These aim at slowing down the growth while the birds are young, in order to achieve better skeletal and cardiovascular development, that allows them to sustain later compensatory growth and to decrease the incidence of late mortality. However, the gain in growth potential and therefore in broiler appetite has been such that the lighting programmes were eventually unable to give satisfactorily control. This has been even truer in conditions where the minimum light duration was 16 hours and light intensity was high. Growth control in this case was obtained by quantitative restriction, with controlled feeding schedules (e.g. 2 x 4-5 hours). The use of such techniques showed an improvement in feed efficiency because feed was better digested compared to adlibitum feeding on the one hand, and because late mortality was decreased on the other hand. Obviously proper implementation of the above solutions requires an extremely good level of management and housing conditions.

The field observations are confirmed by scientists like P. Siege] and I. Nir. More recently, B. Svihus from Norway showed that a pelleted wheat based feed presented as a mash following pellet grinding, reduces feed consumption and increases starch digestibility. This is in line with results from Leciercq (1988) and Plavnick (1995), which indicated that the pelleting effect was essentially explained by a higher feed intake. Svihus's work seems to show that pellet over-consumption leads to impaired feed efficiency because of an alteration in feed intestinal flow regulation. Indeed the gizzard may not be able to properly play its regulatory role through its grinding activity; satiety and thus hormonal or nervous communication with the intestine may be impaired.

The practical consequences of this are that:
- quantitative restriction of broilers improves feed efficiency because of better regulation of intestinal feed flow.
- the gizzard plays a major role in the above feed flow regulation, as long as feed particle size remains coarse.

Mash or Pellet? The Question of Feed presentation, feed factory technology and manufacturing costs

Feed presentation, feed factory technology & manufacturing costs:

Feed Presentation

The energy value of feed or raw materials may vary with type of feed presentation. These have been estimated for the following feed or cereal presentation:
- whole grain cereals added to complementary feed
- cereals or finished feed in mash form
- finished feed in pellet form.

Whole cereal grain is essentially wheat, barley, sorghum (milo) and to some extent whole maize when it is fed to grown birds (more than 30 days of age and / or over 1.2 Kg in bodyweight). Younger birds may be fed broken maize grain.

The energy values are stable, regardless of the presentation of the cereal.

Cereal AMEN (Kcailkg DM)

Wheat energy value is stable regardless of the type of presentation. Whole grain maize shows a 3% lower value, but this may be due to more difficulty for ingestion of grain.

Mixing whole wheat together with feed, either at the farm or at the feed factory, is a frequent practice in Northern Europe. Most of the time, wheat is added at the farm according to broiler performance and sometimes directly at the feed factory, as the feed is loaded onto the feed delivery vehicle. Wheat addition averages between 15 % and 25 %, according to the broiler slaughter weight.

In the Philippines or in Bangladesh, a 50/50 mix of coarsely ground maize and complementary pelleted or mash feed gives very good results.

Cereals/Mash Feed

Mash quality is assessed by the size and uniformity of its particles. A positive correlation between the increase in feed particle size and broiler growth has been demonstrated by several authors, including Nir on 0 to 3 week old chicks, and Lecierc on broilers between 22 to 39 days.

Good uniformity of particle size is essential because birds prefer bigger particles. Thus the dominant birds will quickly eat those bigger cereal particles, while the rest of the birds will eat the finer particles. However, particle selection seems to be balanced by the birds since the cereal/concentrate consumption ratio in free choice eating is very similar to that for the whole feed.

The improvement in performance with feed particle size and uniformity is explained by the lower energy output birds make when they ingest bigger particles. The number of pecks to eat one given feed amount is reduced when particle size increases.

Being grain eaters, birds have a digestive tract designed to quickly ingest large amounts of feed, that are stored in the crop to be 'hydrated" and 'acidified' by lactic acid secretion before going through the proventriculus. In the proventriculus, hydrochloric acid and pepsin and mucus secretions are increased when feed particle size increases. The gizzard carries out feed grinding, feed impregnation and predigestion of the feed by the secretions from the proventriculus, as well as the regulation of feed in-flow and out-flow. This will have an effect upon three digestive flows: from gizzard to proventriculus; from jejunum to duodenum; from rectum to caeca. The intestinal peristaltic motiiity slows down the feed flow, allows better absorpiion of the nutrients by the intestinal viol, and helps to stabiiise the intestinal flora.

Pelleted Feed

The positive effects of pelleting are well documented: higher feed density, no feed ingredient separation, better bacteriological quality, easier ingestion, improved growth and FCR. However, these may vary according to the quality of the raw materials, and that of the grinding and pelleting processes.

The two main physical indicators of pellet quality are:

*   hardness - measured by pellet resistance to breaking when submitted to external pressure
*   durability - measured by the level of "fines" produced during transportation from the feed factory to the farm, and distribution in the feeding system at farm.

Broiler reaction to the above two quality criteria is not easy to assess. In many experiments where better results are obtained with pelleted feed, precise mash characteristics were not given. Indeed, pellets always produce better results when compared to the same fine mash as used to make a good pellet, even more so when the energy level is low. This confirms that the major effect of pelleting dwells in improvement of ingestion.

However, a high energy feed presented either as a coarse mash containing whole grain, or a medium quality pellet because of its fat content, will give very similar results in growth, FCR and fat deposit.

The effect of particle size on growth and consumption of broilers between 21 and 39 days

Broilers may also show high sensitivity to pellet hardness. A hard pellet may be eaten less readily than a softer one, the latter being more likely to yield more fine particles. Still, cereal whole grain ingestion remains easy!

Broiler Performance at 42 & 56 Days

One must be cautious in practice because field conditions are often different compared to experimental conditions. The positive effect of pelleting is essentially due to the improved ingestion. In reality:

- pelleting improves results for low or medium energy feed, which is easy to pellet. The improvement becomes less conspicuous for high-energy feed that is more difficult to pellet and produces more fine particles throughout the transportation and distribution processes

- when high-energy feed is presented as a coarse mash, or as free choice grain + concentrate, the difference between the results is not as significant as when compared to pellet. This may be the wiser choice to reduce the risk of metabolic diseases.

Feed Factory Technology

As shown above, a coarse and uniform mash feed is certainly a good solution for broiler production. Coarse mash that may even be used to produce a pellet is an important factor to regulate feed digestion. Feed particle size depends mainly upon grinding. Two types of grinder are available (cylinder grinder and hammer grinder).

- Grinders with fluted cylinders are not designed for heavy production. They are more sensitive to deterioration by foreign bodies, but they are less power consuming and the feed produced is more uniform in particle size.

Grist Size (MM) According to Roller Distance on Maize

Grinders with hammers are more often used. Grinding is achieved both by contact between feed particles and the hammers and the abrasive effect of the grills, Thus grinding control depends upon two main factors: hammer peripheral speed, and grill mesh size and the percentage of holes. Hammer peripheral speed is a combination of grinder diameter and rotation speed. For one given raw material, the higher the hammer speed, the wider is the distribution range of the feed particles.

Example of calculation:

- rotation speed = 1,500 RPM (Rotations Per Minute)
- grinder diameter = 0.7 m
- peripheral hammer speed = 0.7 x 3.14 x (1,500/60) = 55m/s

If the grinder runs at 3,000 RPM, the speed is 110 m/s. For poultry feed, 55 m/s is the most frequently used speed. Grinders with variable speed allow adapting speed to the raw materials and to the targets for feed particle size.

Grinder Grills: The two important criteria are mesh diameter (from 2 to 10 mm), and the percentage of holes in the grill (from 27 to 52 %). The higher these two values, the higher the average feed particle size and the feed particle size distribution range is.

The dispersion of particle size increases with speed of hammer and size of the grill

The feed particle size and the average distribution range of the feed particle size must be regularly monitored. Excessive variation is a sign of hammer or grill wear.

When blades are worn, the distance between the blade and grill (normally 8mm) is increased. The peripheral feed particle layer therefore becomes thicker and particle ejection is slowed down. The abrasive effect at grill level is increased. Grinder yield diminishes and more fine particles are produced.

In the same manner, worn grills will tend to reject particles back to the grinder instead of letting them out.

For most poultry species, the relevant range for feed particle size is -0.5 to 2 mm. Under 0.5mm, particles are less readily ingested, but this size is essentially composed of vitamins and minerals. Above 2 mm, comprises mostly of the cereals, which may give rise to feed particle selection by the birds.

Grinders with variable speed improve the uniformity of particle size and diminish the amount of particles outside the desired range. Grinders with 55 m/s speed, together with post-grinding sifting to exclude those particles above 3 mm, give good results when working with larger diameter mesh grills to reduce the production of fine particles.

Post Sieving and Recycling

In hot climates, fine mash is not recommended because of its adverse effect on ingestion. Mash should be coarse. The concentrate part of the feed (premix, minerals and proteins) may be presented as a crumble to reduce the amount of fine particles. When this is possible, it is of real interest to give a crumbled starter feed, made from an initially coarse mash.

The % and cumulative % using standard sieves (AFNOR standards)

Manufacturing Costs

Power consumption for grinding and pelleting processes, which represents nearly 80% of the total power requirements, is taken into account.

Average Power Consumption in French Feed Factories

Grinding

Power consumption varies with raw materials and particle size produced

Power consumption and material loss increase with moisture: + 1% moisture =+ 10% more power

Pelleting

This is the highest power consuming operation. Pelleting quality is not easy to assess. Pellet binding must have good adhesion to reduce the production of fine particles during transportation, storage and distribution, yet they must not be too hard in order to avoid possible drops in consumption (Nir). Maize based feed pelleting is more difficult than that of wheat based feed. Several techniques are used to improve pellet binding: finer grinding, the addition of high levels of steam, high temperature (80-85 ºC), possible use of an expander, the level of compression (the compression rate of a dye is calculated by: length/diameter=20). These techniques significantly increase power consumption, yet do they provide real guarantees of improved broiler performance?

In reality, comparing mash and pellet on the basis of the same feed formula is often a biased exercise, as the mash should be finer to obtain a good pellet. To compare pellet with a coarser mash, taking into account the additional grinding and pelleting costs, is more relevant.

Mash or Pellet? The Question of Housing conditions-Human Protection

Housing Conditions - Human Protection

Housing conditions - Consumer's protection: Up until recently, farm results and flock health balance was maintained through the use of antibiotics, growth promoters, and coccidiostats. The recent ban on the use of several products, the limited use of others, the longer withdrawal periods prior to slaughter create a more "ecological' environment where the digestive equilibrium is now achieved by acting on feed and on housing conditions.

Housing Conditions

Monitoring temperature, humidity and ventilation is essential to ensure that the intestine and kidneys function correctly. Too high temperature variations, or air draughts in the house, have an impact on chick feeding behaviour. In low temperature and humid conditions, feed consumption is lower, and both gizzard and intestinal viol development are retarded. Such conditions may be met in tropical climates at night when the temperature goes down to 20-250C and or humidity is high. If houses are not heated properly, light is not sufficient, and ventilation is reduced to try to maintain heat, feed consumption is reduced. During the day, the chick wiii over-consume and drink. If the feed is fine, feed digestive flow is quick and feed assimilation is reduced, which results in diarrhoea (with pasted vent). Very rapidly the flock shows impaired uniformity in growth and feathering even when using antibiotics.

Pellets made using fine grinding and high dye temperature increase the viscosity of the intestinal content. This effect is enhanced when saturated fat (palm oil) is added to the formula. In this case, the osmotic pressure of the intestinal content is increased so much that the resorption of liquids normally occurring in the intestine is blocked. Nutrient assimilation in the duodenum-jejunum loop is impaired, favouring bacteria proliferation in the lower digestive tract. This results in a disruption of the microbial balance leading to development of conforms, salmonella and clostridia. The intestinal microbiological balance can then only be restored by using antibiotics, which today has become an undesirable practice.

The above conclusions can be drawn from many different conditions in different countries and they apply not only to young chicks, but also to adult breeders or grown broilers. Most of the experiments conducted in 'organic" type production conditions, with no use of antibiotics or growth promoters or ingredients from animal origin, show that good monitoring of ventilation, humidity, temperature and more generally hygiene, are key points. One may expect further improvement through the use of vaccines against coccidia, salmonella or clostridia.

Feed: Manufacturing/Formulation

The N° 1 criteria in feed manufacturing is raw material selection at the factory entrance, which must conform to both nutritional and microbiological (bacteria and fungi) criteria.

As explained before, pelleting may be a good method to conform to broiler requirements and to obtain the expected performance. When the pelleting process is achieved under good hygienic conditions, it contributes to lower the bacterial load of the feed (*enterobacteria, salmonella, E. coli,* etc.). However, it may become more risky where the farm technical level is low and the use of antibiotics is strictly limited if not forbidden.

Conversely, a coarse and uniform mash produced from well-controlled raw materials may yield very competitive results. This will be more obvious with higher energy levels. Mash presentation helps steady the digestive process, especially in farms where the technical or health level is not satisfactory,

The above conclusions, together with the current concern for a more 'ecological' production system, have led to changes in the feed manufacturing process: grinding speed reduction to 55m/s (1,500 rpm i.s.o. 3,000 rpm), grit] mesh size increase (4 to 6 mm i.s.o. 3 mm), use of wire-type grills with a higher % of holes and the addition of whole grain (wheat) before or after grinding.

Practically, the ideal balance between an expensive good quality pellet and a cheaper good coarse mash is not so easy to find. There are many different economical solutions in between these two options, in response to the complex equation of raw materials, power costs and housing conditions, etc. In one given environment, the best economical balance may even not be that of top class broiler growth.

Mash or Pellet? We hope we have shown that there is no universal answer and that each set of conditions may lead to a different answer.

The Effect of Low-Density Diets on Broiler Breeder Development and Nutrient Digestibility During the Rearing Period

Low-density diets might help to reduce hunger feeling in restricted-fed broiler breeders. Effects of low-density diets on nutrient digestibility and bird development were studied in Cobb 500 broiler breeder hens from 4 wk of age until the onset of the lay (wk 26).

The experiment included 4 treatments. The control treatment was a normal density diet (ND; 2,600 kcal/kg). Treatments 2 and 3 had a 12 and 23% lower nutrient density than ND (LD12 and LD23, respectively) through inclusion of palm kernel meal, wheat bran, wheat gluten feed, and sunflower seed meal.

Treatment 4 also had a 12% lower nutrient density than ND but included oats and sugar beet pulp (LD12OP).

Total daily intake of energy, digestible lysine, calcium, retainable phosphorus, sodium, and linoleic acid was kept constant in the low-density dietary treatments.

Animal performance and development of the intestinal tract and reproductive tract were determined in addition to digestibility and feed passage rate.

The LD12OP provided a lower AME and nutrient digestibility than calculated and was related to lower live weights.

Birds given LD23 and LD12OP showed a faster ovarian and oviduct development between wk 24 and 26 compared with ND.

At 22 wk of age, LD23 and LD12OP diets resulted in higher empty jejunum and ileum weights. Low-density diets did not affect intestinal tract contents and decreased mean retention time of the contents.

It was concluded that low-density diets can affect live weight and development of digestive and reproductive tracts.

Effect of Maternal and Dietary 25-OH Vitamin D$_3$ on Broiler Production and Immunity

The effects of maternal and dietary 25-OH vitamin D$_3$ (25OHD) on chick immunity and growth were determined. Embryonic mortality was lower in the 25OHD maternal treatment than in the vitamin D$_3$ (D3) treatment. Broiler BW was improved for the broilers fed 25OHD. In vitro chick leukocyte E. coli killing capability was improved with maternal 25OHD. In vitro E. coli killing was increased from immune cell of chicks on maternal and dietary 25OHD. Both maternal and dietary 25-OH D$_3$ improved aspects of broiler production and immune function.

Vitamin D and its metabolites are required for growth, health and bone development in the chick. The natural hepatic production of 25OHD can become impaired, either due to stress such as infection or mycotoxin feed toxicity (Waldenstedt, 2006) or perhaps due to immaturity of enzyme development required for vitamin D absorption in the young chick (Ward, 2004). Providing maternal or dietary 25-OH D_3 to the chick may allow more ready conversion to $1,25(OH)_2D_3$ and therefore may enhance the functions that vitamin D metabolites serve within the body. Previous studies have shown dietary 25OHD to increase BW, improve feed conversion efficiency and increase breast muscle yield (Yarger et a l., 1995) in broilers in comparison with D3.

Some reports indicate that selection for fast growth rates of commercial broilers and turkeys has had a negative impact on the immune response, making modern birds more susceptible to infections than in the past. Vitamin D has been shown to be involved in various aspects of the immune system in humans and other species but its potential function within the chicken has not been well studied.

We have previously shown that 25OHD improved broiler production efficiency and bone quality, and that maternal supplementation improved hatchability and chick early innate immune function (unpublished data). The objectives to the current research were to investigate the potential effects of maternal dietary 25OHD on hatchability as well as maternal and direct dietary supplementation of 25OHD on broiler production and ex vivo innate immune function of the progeny. We hypothesised that maternal 25OHD would support normal broiler breeder production, as well as lead to a more mature innate immune system of the chicks at hatch. In addition, we hypothesised that dietary supplementation of 25OHD would improve broiler production traits.

Materials and Methods

Experimental Diets: Wheat-based basal maternal diets devoid of supplemental vitamin D were formulated for each breeder phase to meet or exceed current primary breeder and NRC (National Research Council, 1994) recommendations. Each basal diet was subdivided and supplemented with either 2,760 IU of dietary vitamin D_3 per kg of feed or 69 µg of dietary 25-OH D_3 (Rovimix D_3 500® or Rovimix HyD® equivalent to 2,760 IU of vitamin D_3 activity, respectively. DSM Nutritional Products Inc., Parsippany, NJ; per kg of feed, was used as the sole source of supplemental vitamin D activity.

Experimental Design and Data Collection: At 23 wk of age, Cobb 500 broiler breeder hens (n=200) were randomly allocated to four floor pens (50 birds per pen; two pens per treatment). In addition, five Ross 308 male broiler breeders were placed in each pen. Birds were weighed and feed allocation adjusted on a weekly basis (separate for males in females) for the average BW of the four pens to maintain the breeder-recommended BW curve. At broiler breeders ages of 31 to 33 weeks (Early), 46 to 48 weeks (Mid) and 61 to 63 weeks (Late), eggs were collected for incubation and hatching. Two complement hatches were done at each broiler breeder age. At hatch, embryonic mortality, hatchability and chick BW were assessed for each maternal dietary treatment group.

Broiler Innate Immune Function and Production Traits: At hatch, chicks were separated based on maternal treatment, and further separated in two additional dietary treatments, resulting in four broiler treatments: Maternal D_3 + Dietary D_3 (DD); Maternal D_3+ Dietary 25OHD (DH); Maternal 25OHD+Dietary D_3 (HD); and Maternal 25OHD+Dietary 25OHD (HH). For one hatch, chicks were place in battery cages and assessed at one and four d post-hatch (n=10 chicks per treatment) were assessed for in-vitro innate immune function. White blood cells phagocytosis of fluorescently-labeled E. coli was analysed (Millet et al., 2007), with modifications to allow analysis by flow cytometry. Oxidative burst was measured using a modified version of that given by He et al., (2003) for analysis by flow cytometry. White blood cell bactericidal (E. coli) capability was measured (Millet et al., 2007). At two weeks of age, five birds per pen were injected intraperitoneally with 3 ml of a 100 µg/ml solution of Salmonella typhimurium lipopolysaccharide (LPS) to simulate an infectious challenge (Korver et al., 1998). Within each pen, five additional chicks were chosen at random to serve as noninjected controls. At 24-hr post-hatch, 10 injected and 10 control birds per treatment were assessed for in-vitro bactericidal capability after an inflammatory immune challenge. For the other hatch per breeder age, chicks were placed in floor pens (four per treatment), assessed for innate immune function as described previously at 1 and 4 d post-hatch as well as for production traits to slaughter weight. Chicks from the Early hatch received a starter ration (0 to 14 d), a grower ration (15 to 27 d) and a finisher ration (28 to 39 d). For the Mid and Late hatches, chicks were fed starter (0 to 10 d), grower (11 to 28 d) and finisher (29 to 41 d) phases. BW on a pen basis was obtained and feed consumption measured for the starter, grower and finisher phases.

Mortality-corrected feed conversion ratios (FCR; g feed/g gain) were calculated.

Statistics: Broiler breeder and hatching egg data were analysed as a one-way ANOVA with diet as the main effect using the Proc Mixed procedure (SAS Institute, 1999). Broiler data were analysed as a 2 X 2 factorial with maternal and dietary treatments as the main effects using the Proc Mixed procedure (SAS Institute, 1999). Means were separated using the LSmeans procedure (SAS Institute, 1999); significance was assessed at P<0.05.

Results and Discussion

Hatchability and Broiler Production Traits: There were no differences in hatch of fertile eggs at any breeder age. Hatchability was on average 92.6, 91.7 and 85.8 % for the Early, Mid and Late hatches respectively. For the Late hatch, the 25OHD breeder treatment group had lower mid (8 to 14 d) embryonic mortality in the vitamin D_3 breeder treatment group (0.67% versus 2.80%, respectively). Chick BW was greater for chicks from 25OHD fed breeders for the Early hatch, however the opposite was found for the Mid and Late hatches. Overall, dietary 25OHD yielded the same hatchability with little difference in embryonic mortality except for the late hatch in which it reduced embryonic mortality. We have previously found that 25OHD significantly reduced embryonic mortality (unpublished data), although the replication was much greater in that study. There were no maternal treatment effects on broiler production traits for the Early hatch, however dietary 25OHD supplementation resulted in greater broiler BW at 21, 27 and 39 d of age. The broilers fed 25OHD tended to eat more feed (significant only for the two weeks feed consumption), but there were no differences in FCR. For the Mid hatch, there were no differences in broiler BW, gain or feed consumption for either the maternal or broiler dietary treatments. However, the maternal vitamin D_3 treatment group had lower FCR for the finisher period than the maternal 25-OH D_3 broilers (1.88 vs 1.94, respectively). For the Late hatch, a nearly significant interaction of maternal and dietary treatment for 10 d BW (P=0.054) and gain (P=0.060) for the starter period. There was an interaction of maternal and dietary treatments in which dietary 25OHD reduced starter period FCR in chicks from the 25OHD maternal treatment but not from the D3 maternal treatment (P=0.038). Dietary 25-OH D_3 resulted in increased broiler BW, although with increased feed consumption for the Early hatch.

The improved FCR during the starter period of the Late hatch may have improved overall production efficiency for those birds.

Innate Immune Function: There were no broiler diet effects on innate immune cell killing of E. coli at either 1 or 4 d post-hatch for the Early, Mid or Late hatches. However, broiler innate immune cell killing of E. coli at 1 d post-hatch for the Early hatch was greater in those chicks from the 25OHD fed breeders (56.1 vs 41.2%, respectively) but not different at 4 d post-hatch. For the Mid hatch, killing of E. coli by immune cells from 25OHD chicks was not different at 1 d post-hatch, but was nearly greater (P=0.063) than the D3 treatment at 4 d post-hatch (32.36 vs 26.36 %, respectively). There were no maternal treatment effects on immune cell killing of E. coli for the Late hatch. After injection with LPS at 2 wk of age, there was no maternal treatment or dietary effects on immune cell E. coli killing for the Mid hatch. However, for the Late hatch, the LPS-injected HH birds had a greater amount of E. coli killing (29.6%), than the LPS injected DD (12.0%) and DH (10.3%) birds; none were different than the HD LPS injected birds (23.4%). These results suggest an improvement in the maturation of the innate immune cell killing capability in young chicks when the hens receive dietary 25OHD. During an inflammatory response, birds from the maternal 25OHD that received dietary 25OHD had among the highest of killing capability.

Both maternal and dietary 25OHD improved aspects of broiler production and innate immune function at all 3 broiler breeder ages. Maternal supplementation decreased embryonic mortality in the Late hatch, and improved early innate immune function of the progeny. Dietary supplementation of 25OHD improved certain aspects of broiler production for the Early and Mid hatches. Therefore 25OHD may be a nutritional means to enhance immune function of poultry without compromising production traits.

Bibliography

Alders, R & Anjos, F & Bagnol, B & Fumo, A & Mata, B & Young, M.: *Controlling Newcastle Disease in Village Chickens,* ACIAR, US, 2002.

Appleby, M.C & Hughes, B.O. & Elson, H.A.: *Poultry Production Systems, Behaviour, Management and Welfare,* CAB International, NY, 1992.

Beck-Chenoweth, H.: *Free-Range Poultry,* Free-Range Poultry Production and Marketing, Creola, Ohio, 2001.

Bell, D.D & Weaver, W.D. & North, M.O.: *Commercial Chicken Meat and Egg Production,* Kluwer, USA, 2001.

Berton, V. and Mudd, D.: *Profitable Poultry: Raising Birds on Pasture,* USDA's Sustainable Agriculture Network (SAN), Washington, DC, 2001.

Blackbourn, David: *The Long Nineteenth Century: A History of Germany,* Oxford University Press, New York, 1998.

Clark, Christopher: *Iron Kingdom: The Rise and Downfall of Prussia,* Cambridge, The Belknap Press of Harvard University Press, Massachusetts, 2006.

Coutts, J.A. & Wilson, G.C.: *Egg Quality Handbook,* NSW DPI, MT, 1990.

Coutts, J.A. & Wilson, G.C. & Fernandez, S. & Rosales, E. & Weber, G. & Hernandez, J.: *Optimum Egg Quality: A Practical Approach,* 5M Publishing, U.K, 2007.

Cramer, C.: *Sustainable Farming Connection: Where Farmers Find and Share Information,* Sustainable Farm Publishing, US, 1997.

Deeming, D.C. & Ferguson, M.W.J.: *Egg Incubation, Its Effects on Embryonic Development in Birds and Reptiles,* Cambridge University Press, UK, 1991.

Dekkers, J.C.M., H.H. Zhao, and RL. Fernando: *Linkage Disequilibrium Mapping in Livestock,* Belo Horizonte, Brazil, 2006.

Dekkers, J.C.M.: *Methods and Strategies for QTL Mapping,* University of Wisconsin, Madison, 2005.

Donald, E. Green: *A History of the Oklahoma State University Division of Agriculture,* Oklahoma State University, Stillwater, 1990.

Fanatico, A.: *Pastured Poultry: A Heifer Project International Case Study Booklet,* Little Rock, AR, 2000.

Frandson, R.D & Wilke, W.L & Fails, A.D.: *Anatomy and Physiology of Farm Animals,* Lippincott, Williams & Wilkins, USA, 2003.

Fraser, AF & Broom, DM.: *Farm Animal Behaviour and Welfare,* CABI, UK, 2004.

French, Jackie: *Jackie French's Chook Book,* Aird Books Pty.Ltd, Australia, 1993.

Gillespie, J.R.: *Modern Livestock and Poultry Production 7th edition,* Thomson Delmar Learning, 2004.

Glatz, P & Bourke, M.: *Beak Trimming Handbook for Egg Producers,* Landlinks, UK, 2006.

Glatz, P.C.: *Poultry Welfare Issues: Beak Trimming,* Nottingham University Press, UK, 2005.

Glatz, P & Ru, Y.: *Developing Free-Range Animal Production Systems,* RIRDC, US, 2004.

Grist, A.: *Poultry Inspection, Anatomy, Physiology and Disease Conditions,* Nottingham University Press, UK, 2006.

Hemsworth, P.H. & Coleman, G.J.: *Human-Animal Interactions: Stockperson-related Issues in the Performance and Welfare of Intensively Handled farm Animals,* CAB International, Wallingford, 2002.

Hunton, P.: *Poultry Production,* Elsevier, 2007.

Huopalahti, R. & Lopez-Fandino, R. & Anton, M. & Schade, R.: *Bioactive Egg Compounds,* Springer, Delhi, 2007.

Hussain, T.; Jilani, G.; Yaseen, M.; Abbas, M.A.: *Effect of Organic Amendments and EM on Crop Production in Pakistan,* SP, Brazil. Pub. USDA. Washington, D.C., 19991.

J. Dennis, Lord: *Encyclopedia of Southern Culture,* University of North Carolina Press, Chapel Hill, 1989.

Jordan, F & Pattison, M. & Alexander, D.: *Poultry Diseases,* Elsevier, 2007.

Jordan, F & Pattison, M & Alexander, D & Faragher, T.: *Poultry Diseases,* WB Saunders, UK, 2002.

Larbier, M.; Leclercq, B.: *Nutrition and Feeding of Poultry,* Nottingham University Press, U.K., 1994.

Lee, A & Foreman, P.: *The Chicken Tractor: The Permaculture Guide to Happy Hens and Healthy Soil-All New Straw Bale Edition,* Good Earth Publications, USA, 2002.

————————: *Day Range Poultry: Every Chicken Owner's Guide to Grazing Gardens and Improving Pastures,* Good Earth Publications, Buena Vista, VA, 2001.

Leeson, S & Summers, J.D.: *Scott's Nutrition of the Chicken,* University Books, UK, 2001.

——————: *Commercial Poultry Nutrition,* University Books, UK, 2005.

——————: *Broiler Breeder Production,* University Books, UK, 2000.

McLelland, J.: *A Colour Atlas of Avian Anatomy,* Wolfe Publishing, London, 1990.

McNab, J.M. & Boorman, K.N.: *Poultry Feedstuffs, Supply Composition and Nutritive Value,* CABI Publishing, UK, 2002.

Mead, G.: *Food Safety Control in the Poultry Industry,* Woodhead Publishing Limited, Abington Hall, Abington, Cambridge, 2006.

Mead, G: *Poultry Meat Processing and Quality,* CRC Press, USA, 2004.

Nicholls, C.: *The Workboot Series: The Story of Eggs,* Kondinin Group, Cloverdale W.A., 2005.

Nicholls, C., Paterson, J.: *The Workboot Series: The Story of Chicken,* Kondinin Group, Cloverdale W.A, 2008.

——————: *The Workboot Series: The Story of Chicken,* Kondinin Group, Cloverdale W.A., 2008.

Nicholls, C.: *The Workboot Series: The Story of Eggs in Australia,* Kondinin Group, Cloverdale W.A., 2005.

Nowland, W.J.: *Modern Poultry Management in Australia,* Rigby, 1978.

Perry, G.: *Welfare of the Laying Hen,* CABI Publishing, UK, 2004.

——————: *Avian Gut Function in Health and Disease,* CABI Publishing, UK, 2007.

Randall, C.J.: *Color Atlas of Diseases and Disorders of the Domestic Fowl and Turkey,* Iowa State University Press, UK, 1991.

Roberts, J.R. & Ball, W.: *Egg Quality Guidelines for the Australian Egg Industry,* AECL Publication, UK, 2004.

Roberts,V.: *Diseases of Free-Range Poultry: Including Hens, Ducks, Geese, Turkeys, Pheasants, Guinea Fowl, Quail and Wild Waterfowl,* Whittet, UK, 2000.

Robinson, F.E & Fasenko, G.M. & Renema, R.A.: *Optimizing Chick Production in Broiler Breeders,* Spotted Cow Press, Ltd, Edmonton, Alberta, Canada, 2003.

Robinson, F.E & Fasenko, G.M. & Renema, R.A.: *New Developments in Reproduction and Incubation of Broiler Chickens,* Spotted Cow Press, Ltd, Edmonton, Alberta, Canada, 2003.

Saif, Y.M.: *Diseases of Poultry,* Iowa State Press, USA, 2003.

Sainsbury, D.: *Poultry Health and Management,* Blackwell Science, US, 2000.

Salatin, J.: *Pastured Poultry Profits,* Polyface, Swoope, Va., 1993.

Salnsbury, D.: *Poultry Health and Management,* Blackwell Scientific, London, U.K., 1992.

Scanes, C.G. & Brant, G. & Ensminger, M.E.: *Poultry Science,* Pearson Prentice Hill, New Jersey, 1992.

Schrijver, R.S. & Koch, G.: *Avian Influenza,* Springer, UK, 2005.

Sim, J. and Sunwoo, H.H.: *The Amazing Egg: Nature's Perfect Functional Food for Health Promotion,* Department of Agricultural, Food and Nutritional Science, University of Alberta, Edmonton, Alberta, Canada, 2006.

Sim, J.S. & Nakai, S & Guenter, W.: *Egg Nutrition and Biotechnology,* CABI Publishing, NY, 2000.

Stadelman, W.J. & Cotterill, O.J.: *Egg Science and Technology,* Food Products Press, Imprint of Haworth Press, New York, London, 1995.

Starck, J.M. & Ricklefs, R.E.: *Avian Growth and Development,* Oxford University Press, New York, 1998.

Stephen F. Strausberg: *From Hills and Hollers: Rise of the Poultry Industry in Arkansas,* Arkansas Agricultural Experiment Station, Fayetteville, 1995.

Tullett, S.G.: *Poultry Science Symposium Number 22.* Butterworth-Heinemann, NJ, 1991.

Watson, R.: *Eggs and Health Promotion,* Iowa State Press, UK, 2002.

Weeks, C & Butterworth, A.: *Measuring and Auditing Broiler Welfare,* CABI, UK, 2004.

Whitehead, C.C.: *Bone Biology and Skeletal Disorders in Poultry,* Carfax Publishing Company, U.K., 1992.

Wiseman, J. & Garnsworthy, P.C.: *Recent Developments in Poultry Nutrition,* University Press, India, 1999.

Yamamoto, T. & Juneja, L.R. & Hatta, H.: *Hen Eggs: Basic and Applied Science,* CRC Press, Delhi, 1996.

Index

❑❑❑